The Making of the Monk

A Monk on World Tour, Volume 1

Sandeep Chavan

Published by SANDEEP CHAVAN, 2025.

THE MAKING OF THE MONK

First edition. August 5, 2025.

ISBN: 979-8231137770

Written by Sandeep Chavan.

Table of Contents

To the seekers—not the followers, not the believers,but those who sit in silence with a storm in their mind,those who ask questions even when answers are offered,those who feel something is missingeven when the world tells them they're complete.

To the ones who have paused mid-successand whispered to themselves,"*There must be more than this...*"

To the doubters who dare to doubt doubt itself.To the rebels who choose reflection over noise.To those who have outgrown comfort and crave clarity.

This book is for you.You are not lost.You are just waking up.

—S.C.

"You were never meant to fit in. You were meant to wake up."— Unknown Monk

Disclaimer

This book is a work of creative nonfiction, philosophical exploration, and narrative insight. While inspired by real-world dilemmas, experiences, and intellectual debates, the characters, events, and dialogues are intentionally designed to provoke thought—not to prescribe ideology or endorse specific beliefs.

The protagonist, Keshav, is a fictional persona used as a narrative voice to reflect the author's ideas, questions, and worldview. Any resemblance to actual persons, living or dead, is purely coincidental unless explicitly stated.

The content within these pages blends science, philosophy, spirituality, and psychology in unconventional ways. It challenges dogma—both scientific and religious—not with the intent to offend, but to **awaken inquiry**.

Readers are encouraged to approach the material with an open mind and personal discernment. This book does not seek followers. It seeks **thinkers**.

All views expressed are those of the author and do not represent any institution, organization, or religious body. No claims are made regarding absolute truth—only the sincere pursuit of alignment with reality.

Preface

This is not a book. It is a rupture.

A rupture in the way we define intelligence, worship knowledge, and mistake performance for purpose. A rupture in the illusion that science alone can explain the soul, or that faith alone can escape the mind.

You are not holding a memoir. You are holding a mirror—one that reflects the collapse of a man who had everything the world celebrates: brilliance, prestige, precision. And yet, standing at the peak of intellect, he saw nothing but emptiness.

Keshav Raghunathan was not meant to be a monk. He was engineered for mastery, not surrender. But intelligence, when awakened, does not obey design. It questions it. It dissolves it. It reclaims the self beneath the system.

This volume is the genesis of that reclamation. It charts the disintegration of a scientist's certainty, the confrontation with intelligence as a living presence, and the radical choice to walk away—not from the world, but from the illusion that the world is enough.

Here, science is not rejected. Faith is not romanticized. Instead, both are stripped to their essence—and reconciled through the one force that transcends them: Intelligence in motion. Consciousness with intent. God, not as a figure, but as the rhythm behind every breath, every question, every act of seeing.

You will not find answers here. You will find the courage to ask better questions.

Because this is not the story of a man becoming a monk. It is the story of a mind refusing to remain programmed.

And if you read it not with agreement, but with awareness— You may begin to see your own reflection in the ripple.

Prologue – A Monk on World Tour

The world has always been divided—science and faith, knowledge and ignorance, intellect and emotion. But is there truly a divide, or have we simply created it?

Some men spend their lives chasing knowledge, believing intelligence is supreme. Others surrender to faith, thinking devotion is wisdom. But few ever realize that the truth is beyond both.

This is the story of a man who had everything—success, prestige, intelligence—but found himself lost in the very knowledge he worshipped.

A man who once believed that faith was a crutch for the weak, that science had all the answers, that intelligence was the pinnacle of human evolution. And yet, standing at the peak of intellect, he saw nothing but emptiness.

What happens when a man questions everything—including his own mind? What happens when a seeker stops searching outward and turns inward?

Keshav Raghunathan was never meant to be a monk.

But some truths do not give you a choice. And some journeys do not end where they begin.

Part I: The Illusion of Success

(Keshav's materialistic world and growing discontent)

1: The Perfect Life?

The rocket launch was a success. The data was flawless. Another breakthrough. Another milestone.

Keshav Raghunathan stood in the control room, watching the live telemetry feed streaming from space. Applause erupted around him—colleagues shook hands, executives smiled, and the director patted his back.

He should have felt pride. Accomplishment. A sense of victory.

Instead, he felt... nothing.

The celebration buzzed around him, but inside, there was only silence. A hollow space that logic couldn't quantify. A feeling he couldn't explain.

The thought came again, uninvited, persistent.

"If science can explain everything, why can't it explain this emptiness?"

A Rising Star in Space Research

Keshav Raghunathan stood in the glass-walled control room, bathed in the glow of hundreds of monitors tracking the progress of their latest satellite launch. The atmosphere was electric—technicians were glued to their screens, analysts relayed real-time data, and a quiet intensity filled the air as the countdown neared completion.

"T-minus ten seconds."

The final moments before a launch always sent a surge of adrenaline through the team. Keshav, however, remained composed. He was used to this. He had seen it countless times before.

"Three... two... one... ignition."

A split second later, the control room erupted into cheers as the live feed displayed the rocket ascending into the dark expanse above. The booster separation was flawless, the telemetry data looked perfect. Another successful launch, another triumph for the company.

Keshav's colleagues patted him on the back, congratulating him on yet another job well done. He responded with a slight nod, offering his signature composed smile. Everyone saw him as a prodigy—the young, brilliant scientist who had carved his path from IIT Madras to one of the most prestigious space research organizations in the United States.

At just thirty-two, Keshav had achieved what most could only dream of. His resume read like a checklist of excellence—top of his class, cutting-edge research in propulsion technology, and a fast-track career in advanced space systems development.

His ideas had revolutionized fuel efficiency in satellite boosters, and his work in autonomous orbital navigation was already making waves in the industry. Senior scientists respected him, executives valued his contributions, and young interns looked up to him as a role model.

For the outside world, Keshav Raghunathan was an unstoppable force, a mind devoted to unraveling the secrets of the universe.

But standing amidst the celebration, watching the rocket disappear into the sky, he felt nothing.

A Life Built on Logic and Precision

Keshav had always been driven by logic, reason, and scientific inquiry. Since childhood, he had been fascinated by the mechanics of the universe—not its philosophy, not its meaning, just the raw, calculable physics behind its workings.

For him, science was the ultimate truth.

Numbers didn't lie. Equations didn't change based on emotion. Everything had an explanation—provided one looked hard enough.

It was this unwavering belief in reason that had carried him through life. While others sought comfort in religion or philosophy, he had found solace in certainty—the ability to predict, calculate, and prove.

Even at IIT Madras, where intellectual discourse often spilled into topics of spirituality and metaphysics, Keshav had remained detached. He saw his peers debating ancient scriptures, pondering consciousness, discussing the Bhagavad Gita's philosophical depth.

"Pointless," he had thought back then.

"Why waste time on things that cannot be measured?"

To him, faith was an excuse for the ignorant, a desperate attempt to explain the unknown. Where others saw mystery, he saw patterns waiting to be solved.

And he had solved plenty.

His research on low-thrust propulsion systems had caught the attention of international space agencies even before he graduated. His Ph.D. work in orbital mechanics had been groundbreaking. By the time he was thirty, he was leading teams on classified space projects, shaping the future of deep-space navigation.

In a world where knowledge was power, Keshav had mastered the art of understanding the universe.

So why, despite all of it, did he feel like something was missing?

The Dream That Became Routine

Every morning, Keshav's routine unfolded with precision, efficiency, and discipline—just like the machines he worked on.

He woke up at 5:30 AM sharp, went for a quick run, followed by black coffee—no sugar, no distractions.

By 7:00 AM, he was at the research facility, buried in calculations, simulations, and project meetings.

Evenings were spent polishing reports, analyzing flight data, and sometimes presenting findings to investors and government officials.

His life was engineered for optimal performance—no wasted time, no unnecessary emotions, no philosophical wanderings.

And yet, in the quiet moments, something gnawed at him.

At dinners with colleagues, he observed them speaking of stocks, promotions, or weekend getaways—all the markers of success.

At corporate events, he watched industry leaders celebrate their achievements, their wealth, their influence.

And in all of it, he saw no real joy.

No sense of fulfillment.

He had everything he had worked for—a career at the forefront of space exploration, a reputation that commanded respect, and the prestige of shaping humanity's future in space.

Yet, at the end of every successful mission, as the celebrations died down, as the screens dimmed, as the noise faded...

A silence remained.

An emptiness that science had no equation for.

The Moment of Disconnect

That evening, Keshav stood by his office window, gazing at the city skyline. The streets below pulsed with movement—people rushing, cars honking, life unfolding at breakneck speed.

Somewhere in that crowd were thousands of dreamers, just like he had been years ago—young engineers, hungry minds, believers in the power of science.

He had been them once.

He had chased knowledge, pursued excellence, and conquered the challenges that stood in his way.

But what now?

Was this all there was to it?

Had he reached the pinnacle, only to find nothing beyond it?

He shook his head, dismissing the thought. It was ridiculous—an absurd distraction. Success wasn't supposed to feel empty.

And yet, it did.

Why?

He had always believed that the pursuit of knowledge was enough. That intelligence alone could bring meaning.

But now, for the first time, a question he had long ignored whispered in the back of his mind:

"Is there more to this world than what science can explain?"

The thought unsettled him. He wasn't one for philosophical nonsense.

And yet, in that moment, as he stared into the vast cityscape, he couldn't shake the growing suspicion that he had spent his life searching in the wrong direction.

That despite everything he knew, despite everything he had built...

He was lost.

The World Through the Lens of Science

Keshav took a slow sip of his coffee, his gaze fixed on the window overlooking the city skyline. The world outside moved with purpose—lights flickered, cars weaved through traffic, and people hurried along, each lost in their own race.

He had always admired the mechanics of movement, the precision of systems—whether it was the orbital trajectory of a satellite or the predictable chaos of an urban landscape. Everything operated on rules, principles, equations. There were no mysteries, only unknown variables waiting to be solved.

This was how he had trained his mind to see the world—not as a mess of emotions, beliefs, or subjective experiences, but as a grand system governed by logic.

Everything had a cause. Every outcome could be explained.

And yet, despite knowing this, despite living by it...

Why did he feel so disconnected?

Faith is for the Uninformed

Keshav had never been a man of faith.

Growing up in India, where religion was woven into the fabric of everyday life, he had been surrounded by people who believed in rituals, temples, prayers, and gods that answered human suffering.

As a child, he had watched his grandmother wake up before sunrise, light incense sticks, and whisper prayers to an unseen force. He had seen his mother fast on holy days, his father visit temples before important business decisions.

But even then, he had felt detached.

"How does praying change anything?" he had once asked his grandmother.

"It's not about changing things. It's about trusting something greater than yourself," she had replied with a smile.

Keshav hadn't understood it then. And as he grew older, he had simply dismissed such beliefs as human coping mechanisms, crutches for those who didn't have the intellect to face reality.

The universe didn't care about faith.

It didn't bend to emotions or prayers.

It followed mathematical precision, absolute laws that required no divine intervention.

Why waste time seeking meaning, when one could seek answers?

Philosophy? Nothing But Mental Gymnastics

If faith was delusion, then philosophy, to Keshav, was nothing but mental gymnastics—endless debates that led nowhere.

At IIT Madras, he had often overheard students discussing Vedanta, existentialism, the nature of consciousness, and the illusions of time.

Some of his closest friends had even entertained the idea that reality itself was subjective, that perception shaped existence, that maybe time wasn't real at all.

Keshav had never entertained such distractions.

"If you can't prove it, it's useless," he would often say.

"But some truths aren't meant to be measured, only realized," a friend had countered once.

He had laughed it off.

Truths that couldn't be measured were not truths at all.

It was precisely this kind of thinking that kept people trapped in outdated ways, clinging to tradition instead of progress.

Philosophy was just another form of religion, except instead of praying, people debated endlessly.

Neither could send a satellite into orbit.

Neither could solve the energy crisis.

Neither had cured diseases, built civilizations, or given mankind the tools to reach beyond its planet.

Science had.

And science was the only thing that mattered.

The Comfort of Logic

For years, this belief had been his anchor, his foundation.

Numbers were absolute. Equations did not change based on personal beliefs.

He had built his entire career, his very identity, on this certainty.

To Keshav, there was a clear distinction between those who questioned reality through science and those who accepted it through faith.

One was progress. The other was stagnation.

One sought answers. The other accepted ignorance.

And yet...

As he sat in his high-rise apartment that evening, staring at the flickering city lights, something felt off.

If logic was enough, if answers were all that mattered...

Why did it still feel like something was missing?

The First Crack in Certainty

The thought was unsettling.

For the first time in his life, he found himself questioning his own convictions.

He closed his eyes and leaned back in his chair, trying to shake the feeling. He had spent years dismissing the search for meaning as human weakness.

But if that were true...

Why did so many intelligent minds—philosophers, poets, thinkers, even some of his own scientific peers—continue to seek something beyond equations?

Why did the greatest minds in history still contemplate existence, consciousness, the nature of the universe?

Surely, they weren't all fools.

Surely, there had to be a reason.

"Everything can be explained."

"So why do people still search for meaning?"

A Moment of Reflection

The question echoed in his mind.

He tried to rationalize it, to dismiss it as a temporary lapse in reasoning.

Maybe it was fatigue. Maybe it was the long hours. Maybe it was just another meaningless thought passing through his mind.

But deep down, he knew better.

This wasn't fatigue.

This wasn't exhaustion.

This was a crack in his certainty.

A question he had ignored for too long.

And the more he tried to push it away, the louder it became.

"What if the answers I seek cannot be found in science alone?"

For the first time in his life, he didn't know.

For the first time in his life, he wasn't sure if he was right.

The Illusion of Control

Keshav had always believed in control.

It was the very foundation of his existence—the ability to predict, measure, and manipulate variables to achieve a desired outcome. Control was order. Control was intelligence in action.

He had spent years perfecting this principle, applying it to his work, his life, his mind.

Science was about control. It turned chaos into structure, the unknown into the known. It gave humans the power to understand the universe, to command the forces of nature, to shape the world according to reason.

Control had taken him from a middle-class home in Chennai to the heart of the world's most advanced space research programs.

Control had earned him respect, recognition, and success.

Control had allowed him to conquer the uncertainties that plagued lesser minds.

Or so he had thought.

The Universe, Reduced to Data

For Keshav, everything had to be quantified.

Every problem had an equation. Every event had a cause. Every outcome was the result of a perfectly calculable sequence of actions.

There was no mystery—only incomplete knowledge.

If something seemed random, it was only because the data set was insufficient. If something appeared unexplainable, it was only because the right formula hadn't been found yet.

This was why science was superior to everything else.

Religion relied on faith.

Philosophy relied on speculation.

Science relied on proof.

And Keshav needed proof for everything.

The Clean, Predictable World of Numbers

At work, his world was built on certainty.

He ran simulations that predicted orbital trajectories with near-zero margin for error.

He developed propulsion systems with equations that left no room for doubt.

He trusted the integrity of the laws of physics—immutable, unchanging, absolute.

His career had conditioned him to believe that nothing was beyond comprehension—just beyond current technology.

Even in his personal life, he lived by precision.

He ate on schedule, exercised on schedule, worked on schedule.

There were no unnecessary emotions.

There was no space for uncertainty.

His apartment was minimalistic—no clutter, no distractions.

His wardrobe was simple—functional, efficient.

His relationships were transactional—networking, not personal entanglements.

Life was optimized. Calculated. Perfected.

Or so it seemed.

The Uncomfortable Truth Beneath the Surface

For all his confidence in numbers and formulas, Keshav could not ignore a growing discomfort.

It wasn't an external problem—it was inside him.

Something was slipping through the cracks of his controlled world. Something he couldn't calculate, measure, or predict.

The nagging emptiness had always been there—he just never had time to confront it.

Now, it was making itself known.

At work, he felt increasingly disconnected. Each successful mission felt like another checkmark, rather than a real achievement.

In conversations, he felt restless. Discussions about stock markets, promotions, and career growth bored him.

At night, he stared at the ceiling, unable to silence his mind. A mind that had always been so sure of itself... now questioned everything.

He tried to suppress it. He distracted himself with more work, longer hours, deeper focus. He told himself it was just fatigue, just a phase, just a meaningless distraction.

But no amount of data, no equation, no calculation could erase the question that refused to leave him:

"If science has all the answers... why do I feel like I'm missing something?"

For the first time in his life, he had encountered an unknown variable he couldn't solve.

And for the first time in his life, he had no answer.

The Hollow Success

The applause had faded. The celebratory emails had been sent. The mission had been a success.

Keshav sat in his office, staring at the numbers on his screen—altitude readings, velocity data, booster efficiency reports. Flawless execution, optimal performance, zero anomalies.

Another achievement.

Another milestone.

Another success that felt like nothing at all.

He leaned back in his chair, rubbing his temples. The emptiness had been growing, spreading through his thoughts like an unshakable shadow. It didn't make sense—he had worked for this his entire life.

This was supposed to be fulfillment.

Then why did it feel so hollow?

The Echo of Empty Conversations

At lunch, he sat with a group of senior scientists, all brilliant minds working on groundbreaking projects. Conversations drifted through the cafeteria—discussions about new contracts, technological advancements, investment portfolios.

A colleague turned to him.

"Keshav, have you thought about applying for the research director role? With your experience, you're a natural fit."

He nodded politely, but something inside him resisted. A few years ago, he would have jumped at the opportunity—a higher position, more influence, the next logical step.

Now, he wasn't sure.

He had spent years climbing the ladder, reaching for the next achievement, the next breakthrough. He had believed that each step forward would bring a sense of completion.

Instead, every milestone felt like a placeholder, a task checked off a never-ending list.

He listened to his colleagues debating the market trends, upcoming promotions, corporate strategy. All highly intelligent minds, discussing things that felt strangely insignificant.

Where was the depth?

Not once did anyone talk about why they were doing what they were doing.

Not once did anyone question if there was more to all of this than just progress.

Keshav stirred his coffee, his mind elsewhere. He wasn't sure when it had started—this sense of detachment, of watching everything from a distance.

Maybe it had always been there, but he had been too focused to notice.

Now, he couldn't ignore it.

The Illusion of Purpose

In the evenings, he would return to his apartment—a sleek, modern space with floor-to-ceiling windows overlooking the city.

A place designed for comfort and efficiency. No distractions. No unnecessary clutter.

And yet, lately, it had started to feel... cold.

Not in temperature, but in essence.

He would pour himself a drink, sit by the window, and watch the world move below. Thousands of people, each following their routines—working, commuting, eating, sleeping.

Just like he was.

Living a life of structure, but was it truly living?

He had always told himself that his work gave him purpose. That his research was pushing the boundaries of human knowledge.

But did it really?

Satellites went up, data came down. Equations were solved, problems were optimized. The system functioned, the world moved forward.

And yet, the core questions of existence remained unchanged.

The Absence of True Joy

There was a time when success had felt euphoric.

When launching a mission had meant pride, excitement, accomplishment.

Now, it felt like a mechanical process—a sequence of steps leading to an outcome that no longer moved him.

He was good at what he did. People respected him. The work had impact.

But where was the joy?

At work, he saw people who lived for the next big project, who felt a rush every time they cracked a difficult problem.

He used to be like that.

But now, every success felt like an echo—loud for a moment, then fading into silence.

There was no lasting fulfillment, only temporary satisfaction, followed by the next goal.

And he was starting to wonder—was he chasing something that didn't exist?

The Question That Wouldn't Leave

One night, he found himself sitting at his desk, unable to focus on his latest project. His mind kept drifting.

He picked up his phone, scrolling through the news—scientific discoveries, economic trends, geopolitical shifts.

Progress, progress, progress.

The world was always moving forward.

But forward to what?

He put the phone down and leaned back.

He had spent his life building a future, but he had never stopped to ask—

For what purpose?

Science had explained so much. But could it explain why he felt empty?

Could it explain why, after everything, something still felt missing?

His rational mind pushed back. This was nonsense.

Emotions were just the brain's biochemical response to stimuli. The feeling of emptiness was just a psychological state, nothing more.

But the explanation felt hollow.

Numbers and equations had always been enough.

Why weren't they enough now?

The Search for More

For the first time in years, he didn't have an answer.

He closed his laptop and looked out the window again.

Maybe... maybe the answer wasn't in more knowledge.

Maybe it wasn't in another project, another mission, another promotion.

Maybe the real question wasn't what's next? but rather—

What am I missing?

The thought unsettled him.

Because for the first time in his life, he realized...

He didn't know.

A Glimpse of Doubt

The city below was alive—pulsing, moving, evolving. From his apartment on the thirty-second floor, Keshav watched as the streets shimmered with headlights, the glass facades of skyscrapers reflecting a million tiny lights, each one representing a life in motion.

People rushing home. People working late. People laughing in cafes, arguing in bars, falling in love, falling apart.

The world never stopped moving.

And yet, standing at the window with a glass of whiskey in his hand, he felt utterly still.

A prisoner in a perfect life.

The Weight of Achievement

He had always imagined this moment differently.

As a teenager, he had dreamed of success—envisioned himself at the pinnacle of his field, leading ambitious space missions, revolutionizing technology.

He had wanted recognition, respect, influence.

And he had it.

At thirty-two, he was among the youngest senior scientists in his organization. His research had shaped the next generation of satellite navigation. His team looked up to him. His name carried weight in the aerospace industry.

By every measure, he was a success.

He had the career, the lifestyle, the intelligence.

So why did it feel like none of it mattered?

He took another sip of whiskey, the burn of alcohol barely registering.

"I should be happy. I have everything I wanted."

The thought felt hollow even as he said it to himself.

Because if that were true, why did it feel like something was missing?

The Silence That Science Couldn't Explain

For years, Keshav had found comfort in numbers, in logic, in certainty.

Success had always been measurable—a set of achievements, equations, and solutions that validated his place in the world.

But now, standing here, away from the equations and algorithms, there was only silence.

A silence that science couldn't fill.

He had spent his entire life searching for answers. But never once had he stopped to ask— Answers to what?

What was he actually solving? What was the purpose of all these calculations, these breakthroughs?

To advance technology? To push the limits of human knowledge?

But for what?

What was the point of progress if he still felt lost?

The thought hit him harder than he expected.

He had spent years believing that intelligence alone was enough. That understanding the universe meant understanding existence.

But here he was, with all the knowledge in the world... and no idea why he felt empty.

The Fear of the Unknown

Keshav was a man of reason.

He had spent his life dismissing doubt as a weakness. A failure of logic.

But tonight, it didn't feel like weakness. It felt like something was trying to break through.

A realization. A whisper in the back of his mind.

Something he had ignored for too long.

Maybe... just maybe... he had been wrong.

Maybe intelligence alone wasn't enough.

Maybe knowledge and meaning weren't the same thing.

The possibility made him uneasy.

Because if that was true... If science didn't have all the answers...Then what else had he been wrong about?

A Lifetime of Avoidance

He had spent years mocking philosophy. Rejecting spirituality. Dismissing faith.

To him, these were the tools of the weak—people who couldn't handle the uncertainty of life, who needed comfort instead of truth.

But tonight, staring into the city, feeling like an outsider in his own life, he wondered—

What if they weren't weak?

What if they had been searching for something he had refused to see?

What if certainty itself was the illusion?

He had always told himself that if something couldn't be measured, calculated, or proven, then it wasn't real.

But then why was this feeling so undeniably real?

He put the whiskey down, suddenly restless.

He needed distraction—work, calculations, something to pull him out of this downward spiral.

He turned back to his desk, opened his laptop.

The screen glowed in the dimly lit apartment.

Telemetry reports. Mission logs. Predictive models.

All things that once gave him control.

But now, staring at the data, it all felt meaningless.

Like numbers that didn't belong to him anymore.

The First Fracture in Certainty

For the first time in his life, Keshav admitted something he had never dared to before.

He didn't have an answer.

For everything he had achieved, for everything he had learned...

He didn't know why it wasn't enough.

And that scared him more than anything.

He shut the laptop and leaned back, staring at the ceiling.

The certainty he had built his life on had cracked.

The world that had once made perfect sense...No longer did.

And the worst part?

He had no idea where that left him.

The Whisper of Change

He exhaled slowly, closing his eyes.

For years, he had seen doubt as a weakness to be conquered.

But maybe doubt was a door.

A door that had always been there... waiting for him to finally open it.

And for the first time in his life,

He wasn't sure if he wanted to walk through.

2: The Noise of the World

The hum of conversations filled the executive lounge—a constant, rhythmic murmur of voices, deals, and ambitions. Laughter punctuated the air, but it was the kind of laughter that didn't reach the eyes.

Keshav sat at a polished mahogany table, his coffee untouched, his mind elsewhere.

Around him, some of the most brilliant minds in aerospace engineering were engaged in discussion—not about science, exploration, or the pursuit of knowledge, but about funding projections, stock performance, and corporate politics.

"We just closed another $200 million contract—our valuation is skyrocketing."

"Did you hear? The new director is playing favorites—if you want that promotion, better align with him."

"The government is pumping more into private space research next year—we need to position ourselves for maximum stake."

Keshav listened, but didn't participate.

A few years ago, he might have joined the conversation, played the game.

Now, he was only observing.

Where was the passion? The curiosity? The hunger to understand the universe?

These were men and women who had dedicated their lives to science.

Yet, as he looked around, he saw exhaustion behind their smiles, stress behind their successes.

Intelligence had made them powerful.

But it hadn't made them happy.

And Keshav couldn't ignore the question forming in his mind—

"If intelligence is supreme, why does it make people so unhappy?"

The Unending Race

The day began like every other—fast, urgent, relentless.

Keshav stepped out of his car and into the towering glass structure of the aerospace facility. The sun was just beginning to rise, casting a golden glow over the city, but inside these walls, the world had already been awake for hours. There was no sunrise here—only deadlines, reports, and the race to stay ahead.

As he walked through the entrance, he could already hear it—the low hum of a machine that never stopped running.

Monitors flashing with data. Keyboards clacking in rapid succession. Engineers moving between conference rooms, their heads buried in tablets. A team huddled around a screen, debating last-minute adjustments to a satellite model.

Everything moved at a pace that left no room for hesitation.

The frenzy of progress was intoxicating to some. To Keshav, it was beginning to feel like noise.

Meetings, Deadlines, and the Corporate Pulse

He barely had time to set his bag down before the first meeting of the day began.

A room filled with executives, engineers, and research heads—all focused on one thing: results.

"We need updated trajectory models for the new satellite launch—margin of error needs to be below 0.002%."

"Government funding is shifting priorities. We need to reframe our research to match their objectives."

"Our competitors just announced a new deep-space propulsion model. We need to move faster."

Faster. More efficient. More innovative.

The pace never slowed.

Keshav sat through it all, offering his insights, approving new calculations, adjusting project timelines. He was good at this—at solving problems, at making decisions.

But somewhere in the back of his mind, a quiet voice whispered:

"Is this all it is? A never-ending cycle of meetings and missions, contracts and competition?"

He pushed the thought aside. There was no time for reflection.

The Industry That Never Sleeps

By mid-morning, he had moved from technical briefings to investor calls.

Gone were the conversations about scientific exploration, about pushing the frontiers of knowledge.

Now, the language was different.

"What are the projected returns on this technology over the next five years?"

"How does this position us against our competitors?"

"Will this project increase shareholder confidence?"

The research that once thrilled him had become a product. A commodity. A thing to be sold, packaged, and marketed to the highest bidder.

It was no longer about understanding the universe.

It was about securing the next round of funding, the next corporate partnership, the next expansion.

The business of science had overtaken the purpose of science.

And yet, everyone around him seemed to accept this as normal.

The Relentless Pursuit of Progress

Progress.

It was the word everyone used to justify the chaos.

The late nights. The impossible deadlines. The constant pressure to innovate.

"We are building the future." "We are shaping the next era of space exploration." "We are advancing human knowledge."

But Keshav had begun to wonder—to what end?

Progress toward what?

What happened when they reached the next milestone?

Would there be fulfillment? Satisfaction? Or just another bigger, faster, more complex project waiting on the horizon?

He had spent his life chasing progress, believing that each breakthrough would bring him closer to something meaningful.

But lately, it all felt like a treadmill—running faster, moving forward, but going nowhere.

Everyone is Running—But Why?

That afternoon, he stepped out onto the terrace for a moment of quiet.

Below, the city moved in a synchronized chaos.

Cars honking. Pedestrians rushing. Office buildings filled with people typing, talking, selling, building.

Everyone was moving toward something.

A promotion. A bigger paycheck. A better title. A new business deal.

A constant pursuit of more.

He thought of his colleagues—scientists, engineers, some of the most intelligent minds in the world.

All running, achieving, reaching for the next level.

But for what?

Did any of them stop to ask where it all led?

Or was this just a race with no finish line?

The Uncomfortable Truth

For years, Keshav had loved the race.

He had thrived on pressure, on solving complex problems, on pushing boundaries.

He had believed that progress was the highest pursuit.

But now, standing on that terrace, looking at a world that never stopped moving, he wasn't so sure.

What if progress wasn't enough?

What if the pursuit itself was flawed?

What if, in chasing something greater, he had missed the point entirely?

The thought unsettled him.

Because it wasn't just about the industry, the system, or the corporate machine.

It was about him.

For the first time, he asked himself a question he never had before.

"What am I running toward?"

And for the first time in his life...

He didn't have an answer.

Trapped in the System

Keshav walked back into the control center, the air thick with tension.

Rows of engineers sat hunched over their screens, analyzing real-time telemetry, their eyes darting between data points. A group of senior scientists stood in the corner, engaged in a hushed yet intense debate over a technical anomaly. Managers moved between workstations, their voices laced with urgency.

It was a familiar sight.

A launch window approaching. A system needing last-minute refinements. A room full of brilliant minds, working under relentless pressure.

To an outsider, it would seem inspiring—some of the world's most intelligent people, collaborating at the peak of human innovation.

But to Keshav, it was something else entirely.

It was exhaustion.

The Cycle of Stress

He could see it in their faces—fatigue hidden behind caffeine-fueled energy, stress concealed beneath a forced sense of urgency.

Some had been here for fifteen hours straight. Others had barely eaten.

A few were so lost in the complexity of their work that they didn't even realize how drained they looked.

These were people who had once been driven by curiosity, by passion, by the desire to push the limits of human knowledge.

Now, they were just trying to keep up.

Trying to meet deadlines. Trying to secure funding. Trying to survive the system that was supposed to empower them.

Slaves to Numbers, Not to Knowledge

Keshav took his seat at the review panel, listening to the team lead go over the latest performance reports.

"Propulsion efficiency is within expected parameters, but we may need to optimize thrust-to-weight ratio before final approval."

"We need to improve cost projections—investors want to see at least a 12% reduction in material expenses before they approve the next phase."

"We've got to streamline our process to align with next quarter's budget constraints."

Keshav scanned the room, watching how his colleagues absorbed this information.

Some nodded mechanically, scribbling notes. Others ran quick recalculations, adjusting figures to meet budget expectations. No one questioned the purpose behind it.

This wasn't about knowledge anymore.

It wasn't about discovery, innovation, or pushing the boundaries of science.

It was about metrics.

Everything reduced to percentages, performance indicators, cost-benefit analyses.

And the people who had once dreamed of exploring the unknown were now just managing numbers.

The Business of Science

Later that afternoon, Keshav attended a high-level corporate briefing.

Executives and department heads gathered in a sleek, modern conference room. The discussion was not about scientific breakthroughs or the future of space exploration.

It was about profits, partnerships, and market competition.

"Our research team has been approached for a joint venture—there's potential for a multi-billion-dollar investment, but they want exclusivity on propulsion data."

"Can we afford to allocate more resources to deep-space navigation? Or do we shift focus to commercial satellite deployment? That's where the real money is."

"We need to start thinking about patents and licensing—we can't just build, we need to own the market."

Keshav sat silently, observing.

He had known, of course, that research required funding. That corporate partnerships were necessary to sustain large-scale projects.

But somewhere along the way, something had shifted.

This wasn't just science adapting to business.

This was science becoming business.

Decisions were being made not on the basis of what would advance human knowledge, but on what would generate the highest returns.

And the people who had once dedicated their lives to understanding the cosmos were now discussing profit margins.

The Irony of Intelligence

Keshav left the meeting with a strange heaviness in his chest.

He had spent his entire life believing in intelligence—that the pursuit of knowledge was what set humans apart, that scientific discovery was the most noble and purposeful endeavor.

But now, he saw the truth.

Intelligence was not liberating these people. It was enslaving them.

These were some of the smartest people on the planet.

Yet, they were trapped in the very system they had helped build.

Not free to explore, but forced to justify their work in financial terms.

Not leading progress, but adapting to market trends.

Science had once been about seeking the truth. Now, it was about securing the next round of funding.

And for the first time, Keshav wondered—

"Is this the price of progress?"

The Slow Death of Passion

That evening, Keshav met an old colleague for dinner.

A man who had once been one of the most passionate scientists he had ever known.

Someone who had pushed boundaries, questioned everything, chased knowledge for its own sake.

Now, he looked tired.

"You're lucky, Keshav," he said, sipping his whiskey. *"You still believe in this work. Some of us... we've just learned to play the game."*

Keshav frowned. *"What game?"*

The man chuckled, shaking his head.

"The game of knowing how much to care. Care too little, and you won't get anywhere. Care too much, and it'll break you."

He leaned back, sighing. *"We all started out chasing knowledge. But in the end, we're just chasing deadlines."*

Keshav left the restaurant that night with a new thought.

What if this entire system was broken?

What if the smartest minds in the world were wasting their brilliance—not on discovery, but on survival?

What if the pursuit of knowledge had become just another industry?

The Question That Wouldn't Leave

The next morning, Keshav sat at his desk, staring at an untouched report.

A question had taken root in his mind, and no matter how hard he tried to push it away, it kept returning.

"If we are building for business, not for knowledge... then what is the purpose of any of this?"

He looked around the office.

Brilliant people, working relentlessly. Deadlines driving every decision. Science being measured in profit, not in progress.

For the first time, Keshav didn't want to be here.

And for the first time, he realized—he didn't have to be.

The Language of Money and Power

Keshav sat in the boardroom, a sleek, glass-walled chamber perched high above the city. Outside, the skyline stretched endlessly—steel and concrete monuments to human ambition. Inside, the air was thick with the weight of decisions that had nothing to do with science.

He had attended hundreds of these meetings before, but lately, they felt different.

He wasn't sure if the meetings had changed, or if he had.

The conversations around him had nothing to do with discovery, exploration, or pushing the boundaries of human knowledge.

Instead, the discussions were about budgets, investors, patents, and market competition.

"We need to shift funding into projects that guarantee higher returns—government contracts are more secure than deep-space exploration."

"The private sector is investing heavily in satellite-based surveillance—we should be positioning ourselves in that market."

"Cost overruns in the propulsion department need to be addressed before the next investor review."

Numbers. Projections. Risk assessments.

This was not science.

This was business.

And Keshav realized, with unsettling clarity, that scientists weren't leading the future anymore—executives were.

The Corporate Machine

A director at the far end of the table pulled up a financial breakdown on the massive display screen.

Colored graphs, profit margins, long-term investment strategies—everything quantified, optimized, structured for maximum return.

"Our research trajectory needs to align with revenue potential," the director explained. *"We cannot afford to invest in long-term projects that do not have clear profitability models."*

Keshav sat back in his chair, watching as his colleagues nodded in agreement.

Gone were the days when research was driven by curiosity and the pursuit of knowledge.

Now, even discovery had to justify itself in economic terms.

He had always known that funding was important—but when had it become the primary driver of innovation?

The Politics of Science

Keshav had never cared much for corporate politics.

He had always assumed that intelligence alone was enough—that if you worked hard, solved complex problems, and contributed meaningful ideas, you would succeed.

But the reality of the corporate world had little to do with intelligence.

Success wasn't just about who was the smartest—it was about who could navigate power structures the best.

It was about alliances, influence, and strategic positioning.

Who had the backing of the board? Who was aligned with the leadership's vision? Who could secure the biggest funding deals?

And who was willing to play the game?

Keshav saw it happening around him—brilliant scientists being overlooked for promotions because they lacked "leadership presence," while those who knew how to speak the language of executives climbed higher.

It wasn't about who had the best ideas. It was about who had the most influence.

He realized something that should have been obvious—scientists weren't in control of science anymore.

The decisions that shaped the future of technology, space exploration, and human progress were being made by business strategists, financial analysts, and corporate executives.

And the irony was inescapable.

The smartest minds in the world were still playing by the same materialistic rules as everyone else.

Research as a Commodity

Funding dictated everything.

Not potential. Not the pursuit of knowledge. Not the impact on humanity.

Only profitability.

Keshav had seen it happen time and time again—promising projects getting shelved because they didn't offer a clear revenue model.

Scientists forced to align their research with corporate interests rather than their own intellectual curiosity.

Even when breakthrough discoveries were made, the focus immediately shifted to monetization, patents, and exclusive licensing deals.

"If we can commercialize this, we can dominate the market for at least a decade."

"Can we make the research proprietary? We need to ensure competitors don't gain access."

"If the military funds this, we could secure indefinite contracts—pure scientific exploration isn't profitable enough."

Keshav had spent his life believing that knowledge was the highest pursuit.

That truth had no price tag.

But now he saw that even knowledge had been turned into a product.

The Illusion of Progress

The company liked to talk about pushing the boundaries of science.

They used words like exploration, discovery, advancement—but only when it aligned with their business strategy.

Progress was no longer about expanding human understanding.

It was about expanding market share.

And the people at the top weren't scientists, engineers, or thinkers.

They were investors, shareholders, and business executives who saw science not as a calling, but as a tool for financial growth.

Keshav saw the hypocrisy of it all—the people who spoke of discovery were the same ones ensuring that true discovery was always secondary to economic gain.

A Quiet Realization

After the meeting, Keshav walked down the hallway, feeling an unfamiliar weight pressing against him.

He had spent years believing in the purity of science.

He had convinced himself that he was part of something greater, a movement dedicated to understanding the universe, not selling it.

But now, he wasn't so sure.

He reached the balcony, looking out over the city.

For years, he had thought science was the path to truth.

But what if truth had nothing to do with it anymore?

The Question That Wouldn't Go Away

He had spent his life believing that intelligence was above the material world.

That scientists were different from businessmen, that researchers were different from politicians.

But now he saw the truth—they were not different at all.

The world of science was just as materialistic, just as political, just as driven by power as anything else.

It wasn't about knowledge anymore.

It was about who controlled knowledge.

And the more he thought about it, the clearer it became—

"Maybe I don't belong here anymore."

The Absence of Joy

The evening air was heavy with the hum of conversation as Keshav sat across from his colleagues at an upscale restaurant. It was a celebratory dinner—another successful mission, another milestone reached. A night to recognize the brilliance of the team.

Waiters moved effortlessly between tables, the soft clinking of glasses and muted jazz creating an illusion of ease, of celebration.

But Keshav wasn't celebrating. He was watching.

For years, he had admired the people around him—scientists, engineers, strategists, minds that had shaped the future of technology and space exploration.

Now, he saw them differently.

Not as pioneers of knowledge, but as burned-out intellectuals, tired men and women who had given everything to their work and had little left for themselves.

They had respect, wealth, achievements—yet they carried the weight of exhaustion in their eyes.

And for the first time, Keshav wondered—is this what success looks like?

The Cost of Intelligence

He listened as the conversation drifted between corporate shifts, promotions, and industry politics.

"I should be taking time off, but there's too much at stake with the next project."

"I haven't had a full weekend to myself in over a year."

"I love the work, but damn, sometimes I wonder if it's all worth it."

Laughter followed these admissions, but it wasn't the kind of laughter that carried warmth.

It was the kind that masked exhaustion.

The kind that covered up something deeper.

Keshav had heard these same lines for years.

He had even said them himself.

But tonight, they felt different.

He had spent his life believing that intelligence was the highest pursuit. That if you were brilliant, successful, and accomplished, you would naturally be happy.

But now, as he looked at the people who had achieved everything—he realized something terrifying.

None of them were happy.

A Room Full of Achievers, A Room Full of Emptiness

These were men and women who had reached the pinnacle of intellectual achievement.

They had devoted decades to solving problems that shaped the future of space exploration.

They had contributed to some of the most advanced technological breakthroughs in history.

Yet, they looked exhausted.

They spoke of their work with pride—but rarely with joy.

There was no passion left in their voices.

Their laughter was short, their drinks stronger than necessary.

Keshav studied them carefully.

If these people—brilliant, driven, successful—weren't happy, then who was?

And if achievement didn't bring fulfillment, what did?

Success Without Purpose

For years, Keshav had believed that the pursuit of knowledge was enough.

That the act of learning, building, discovering was in itself a source of meaning.

But here he was, sitting among some of the greatest minds in his field, and all he saw was detachment.

The higher they climbed, the less connected they seemed to anything real.

They weren't just detached from society, from their families, from their health.

They were detached from themselves.

From the very thing that had once made them fall in love with science.

Somewhere along the way, in the relentless pursuit of progress, they had lost the ability to feel joy.

The Illusion of the Next Achievement

Keshav leaned back in his chair, listening.

One of his colleagues was speaking about an upcoming leadership position—a major promotion, the kind that only came once in a career.

"I wasn't even sure if I wanted it," he admitted, swirling the whiskey in his glass. *"But hell, after working this hard, I guess I have to take it."*

Another nodded. *"You don't work this hard just to say no."*

Keshav felt a chill run through him.

They weren't chasing happiness.

They were chasing momentum.

Success wasn't something they enjoyed.

It was something they endured.

And once they reached one milestone, they didn't stop to appreciate it—they immediately set their sights on the next one.

Because slowing down meant confronting a truth they didn't want to face.

That no matter how much they achieved, it would never be enough.

What Are We Running Toward?

On the drive home, Keshav was quiet.

His mind replayed the night, the words, the faces, the laughter that wasn't really laughter.

For years, he had embraced the race—believing that progress was everything, that the pursuit of knowledge was the most meaningful thing in life.

But what if he had been wrong?

What if the problem wasn't just in the corporate world, but in the very structure of ambition itself?

What if the smartest people in the world had been running in circles?

Pushing forward, never stopping, never questioning—not because they were sure of where they were going, but because they were afraid of what would happen if they stopped.

And now, for the first time, he was afraid too.

Because what if there was nothing at the end of this race?

A Truth He Could No Longer Ignore

Keshav sat in his apartment, staring at the skyline.

He had spent his life believing that knowledge was the answer.

But tonight, he had seen something that no equation could explain.

Intelligence had given these people everything—power, wealth, influence.

But it had taken something too.

Something essential.

Something he couldn't quite define.

He took a deep breath, exhaling slowly.

For the first time in years, he didn't feel certain about anything.

And that terrified him.

Because if intelligence wasn't enough...

Then what was?

The Question That Won't Go Away

The city stretched out before him, a vast ocean of lights flickering against the night sky.

From the window of his high-rise apartment, Keshav could see hundreds of lives unfolding below. Office workers making their way home, taxis weaving through congested streets, neon signs casting their glow onto busy sidewalks.

A world constantly in motion.

A world full of people—chasing, striving, reaching for something.

And yet, the thought wouldn't leave him.

"If intelligence is the highest power, why does it make people so unhappy?"

He had spent his life believing that knowledge was the answer to everything. That the more one understood, the closer they came to truth, to purpose, to meaning.

But tonight, that belief felt fragile.

Because if it were true, then why did the most brilliant people he knew seem so lost?

Why did their intelligence fail to bring them fulfillment?

Why did they, despite knowing more than anyone else, still feel incomplete?

And more than anything—why was he starting to feel the same way?

The Contradiction He Couldn't Ignore

For years, Keshav had equated intelligence with superiority.

If you were intelligent, you were in control. If you were intelligent, you had the answers. If you were intelligent, you didn't need anything else.

But now, for the first time, he was questioning it all.

Because he had seen it with his own eyes—people who had reached the highest levels of knowledge, yet still struggled with the same doubts, the same anxieties, the same discontent as anyone else.

These were men and women who had sent missions beyond Earth, designed technology that reshaped industries, developed theories that challenged the boundaries of human understanding.

And yet, they weren't happy.

They were stressed, unfulfilled, caught in the same endless cycle of deadlines, promotions, and achievements.

And if that was true for them...

Then what did that mean for him?

The Realization He Didn't Want to Have

Keshav leaned against the window, his hands in his pockets, his mind a battlefield of logic and doubt.

Maybe he had misunderstood everything.

Maybe intelligence wasn't enough.

Maybe knowledge alone didn't bring meaning.

The thought made him uneasy.

For so long, he had dismissed philosophy, faith, and introspection as distractions. Tools for the weak-minded, crutches for those unwilling to face reality.

But now, as he stood here—accomplished, respected, and yet deeply unsettled—he wondered if he had been blind to something all along.

Something bigger than equations and algorithms. Something bigger than knowledge itself.

But what was it?

The Doubt That Refused to Die

He tried to push the thought away.

Tried to rationalize it.

"This is just stress. Fatigue. A temporary phase."

But deep down, he knew better.

This wasn't fatigue. This wasn't stress. This was a realization creeping in—one he could no longer ignore.

He was not fulfilled.

Not despite his success, but because of it.

He had achieved everything he was supposed to.

And yet, it wasn't enough.

Which meant maybe it was never supposed to be.

The Fear of What Comes Next

For years, he had been sure of his path.

Sure of who he was. Sure of where he was going. Sure of what mattered.

But tonight, for the first time, he felt completely lost.

Because if intelligence wasn't the answer...

Then what was?

And if he had spent his life chasing something that could never bring him peace...

Then where did that leave him now?

He closed his eyes, exhaling slowly.

For the first time in years, he had no answers.

Only questions.

And worse—

For the first time, he wasn't sure if he wanted to find out the truth.

3: The First Crack

A Simple Question

The hallway was empty, the usual hum of voices replaced by the quiet whir of ventilation systems.

It was late. Most of the staff had left hours ago, but Keshav remained, buried in work, trying to outrun the silence that had begun creeping into his thoughts.

He stepped out of his office for a break, heading toward the vending machine for another caffeine fix. As he reached for his coffee, a voice broke the stillness.

"Working late again, sir?"

Keshav turned to see an older janitor, a man he had seen countless times but never really noticed. Grey hair, kind eyes, the relaxed ease of someone who had lived long enough to see patterns repeat.

Keshav nodded, offering a polite but distracted smile. He wasn't in the mood for small talk.

But then, the janitor spoke again.

"You study the universe, sir... but do you study yourself?"

Keshav frowned, caught off guard.

"What?" he asked, half-laughing. *"I study reality. The universe. The things that actually matter."*

The old man just smiled.

"And where do you fit into all of that?"

Before Keshav could respond, the janitor simply nodded, picked up his bucket, and walked away, leaving Keshav standing there—coffee in hand, thoughts in turmoil.

It was just a question.

A simple, meaningless question.

Yet, as Keshav returned to his office, he couldn't shake the feeling that something had shifted.

And for the first time, his mind—always looking outward—turned inward.

The Unlikely Encounter

The office was silent, the usual chatter of scientists and engineers replaced by the distant hum of servers and the rhythmic tapping of a keyboard. Keshav sat at his desk, staring at rows of complex equations on his screen, but his mind wasn't fully present.

It had been happening more often lately—his ability to focus was slipping. No matter how much he immersed himself in data, in calculations, in the science he had devoted his life to, there was something... off.

Somewhere, just beneath the surface of his thoughts, a question lingered. A question he didn't want to face.

To shake it off, he leaned back and rubbed his eyes. It was late. Most of the facility had cleared out hours ago, but staying behind had become his habit. He had convinced himself that more work meant more control—that if he kept himself occupied, the creeping doubts would fade.

But they weren't fading.

Sighing, he stood up and stretched. Maybe a break would help.

The Unexpected Conversation

As he made his way down the dimly lit corridor, he passed the glass walls of the main research lab. The monitors inside glowed in the dark, displaying real-time data from a recently launched satellite. Somewhere in deep space, a machine built by human hands was orbiting the void, gathering information that would push the boundaries of knowledge.

It should have felt exhilarating. Once, it had.

Now, it just felt... routine.

He turned a corner, heading toward the break area where a vending machine stood next to a small seating space. As he reached for his usual black coffee, a voice interrupted the quiet.

"Working late again, sir?"

Keshav turned and saw the janitor—a man he had seen countless times but had never really noticed.

He was an older man, likely in his sixties, with a slow but steady posture. His uniform was worn, his hands rough from years of labor, yet his face carried the kind of calmness that Keshav found unsettling.

"Yes," Keshav replied, offering a polite but distracted nod. *"Always something to work on."*

The janitor nodded, placing his mop back into the bucket.

"Smart people always work late," he said casually.

Keshav smirked, unsure if it was a compliment or just an observation.

"Somebody has to keep things running," he said.

The janitor chuckled. *"Oh, I don't mean the facility, sir. I mean up here."* He tapped his temple. *"Smart people are always trying to solve things. Even when they're not working, their minds never stop."*

Keshav hesitated for a moment before responding. *"That's the nature of progress. We ask questions and find answers. That's what moves the world forward."*

The janitor gave him a thoughtful look.

"You study the universe, sir... but do you study yourself?"

The words were simple. Too simple.

Keshav almost laughed.

"What?" he asked, raising an eyebrow. *"I study reality. The universe. The things that actually matter."*

The janitor just smiled, unmoved by his answer.

"And where do you fit into all of that?"

Keshav opened his mouth to respond, but nothing came out.

Before he could think of an answer, the janitor simply nodded, picked up his bucket, and continued down the hallway, humming an old tune as he disappeared around the corner.

Brushing It Off

Keshav stood there for a moment, coffee in hand, staring at the empty hallway.

What kind of question was that?

It was nonsense—something a philosopher or a mystic might ask, not something relevant to the real world.

He shook his head, brushing it off.

"Old man wisdom," he muttered under his breath as he walked back to his office.

Yet, as he sat back down at his desk, the question wouldn't leave him.

"You study the universe, but do you study yourself?"

It echoed in his mind, like a tune he couldn't stop humming.

Why did it bother him so much?

Maybe it was because he had never thought about it before.

He had always believed in looking outward—toward data, discoveries, the vast unknown of space.

He had never once turned his focus inward.

Because why would he?

There was nothing to discover there.

Was there?

The First Crack

He forced himself to dive back into his work.

Equations. Trajectories. Satellite diagnostics.

Numbers always made sense. They were pure, logical, absolute.

But tonight, they felt different.

For the first time, they felt... detached.

Like he was looking at them, but not connecting with them.

Like they were part of a world he had built for himself, but no longer fully belonged to.

He shook his head and closed his laptop.

Maybe he just needed sleep.

Maybe he was overthinking.

But deep down, he knew the truth.

A question had been asked.

And no equation could answer it.

The Beginning of Introspection

As he lay in bed that night, the city lights casting faint shadows on his ceiling, he tried to push the thought away.

"I study the universe, but do I study myself?"

He closed his eyes.

For years, he had been so sure of the path he was on.

So sure that knowledge was the highest pursuit. So sure that science held the answers to everything. So sure that introspection was a waste of time.

But now, doubt had crept in.

And no matter how much he tried to fight it, he couldn't help but feel like something had shifted.

For the first time in his life...

He wasn't certain anymore.

Dismissing the Question

Keshav shook his head as he walked back to his office, the janitor's words still lingering in his mind.

"You study the universe, but do you study yourself?"

It was absurd.

Meaningless.

Something an old man with too much time and not enough education might say.

He chuckled to himself, dismissing the thought as easily as he would dismiss a flawed hypothesis.

His entire life had been built on logic, observation, and empirical evidence. He had no patience for vague philosophical musings that led nowhere.

"The universe is far more complex and fascinating than human emotions," he muttered under his breath, as if saying it out loud would make it more real.

The stars, galaxies, black holes, and the very fabric of spacetime—these were the things worth studying.

Not introspection. Not self-awareness. Not pointless existential questions.

And certainly not the words of a janitor.

Yet, as much as he tried to push it away, the question refused to leave him.

A Mind Conditioned for Logic

Keshav had spent years training his mind to filter out unnecessary noise.

When he was a student at IIT Madras, he had no time for distractions.

While his classmates engaged in long discussions about consciousness, philosophy, and the meaning of life, he focused on what could be measured and proven.

Some of them had even indulged in spiritual texts, quoting the Bhagavad Gita and Vedanta, discussing the nature of the self.

He had always laughed at them.

"You're wasting your time," he had told them. *"The world doesn't need more poets. It needs engineers, scientists, and problem-solvers."*

And now, after all these years, a janitor had asked him a question that had momentarily short-circuited his logical mind.

It frustrated him.

Why had he hesitated? Why had his mind gone blank instead of coming up with a sharp, scientific response?

He should have shut the conversation down immediately. He should have dismissed it outright. He should have...

But he hadn't.

Something about the simplicity of the question had unsettled him.

A Forced Laugh

Sitting at his desk, Keshav let out a small laugh, shaking his head.

It was ridiculous.

He had spent the last fifteen minutes thinking about a random remark from a janitor.

Of all the things he could be focusing on—orbital mechanics, propulsion efficiency, upcoming project deadlines—he was sitting here debating a meaningless philosophical question.

"Maybe I'm just tired," he reasoned. *"Too much work. Too little sleep."*

Yes. That had to be it.

The human brain—even an intelligent one—was still susceptible to fatigue-induced distraction.

"That's all this is," he told himself. *"A distraction."*

But deep down, he wasn't convinced.

The Janitor's Smile

Keshav leaned back in his chair, replaying the conversation in his head.

He had expected the janitor to look impressed by his response.

To nod and agree with the words of an educated man.

To acknowledge that science was the highest pursuit and that philosophy was just speculation.

But the janitor had done none of that.

He had simply smiled.

A knowing smile.

Like he had heard this exact argument before.

Like he had seen smarter men than Keshav say the same thing—and still come back later, questioning everything.

That smile bothered him more than the words.

It wasn't the smile of someone who had lost a debate.

It was the smile of someone who knew something Keshav didn't.

And that was deeply unsettling.

Trying to Move On

Keshav forced himself to return to work, pulling up the latest research reports.

Telemetry analysis. Fuel consumption data. Predictive modeling for trajectory shifts.

Things that actually mattered.

For a while, he managed to lose himself in numbers, in the clean precision of equations.

But then, in the quiet moments between calculations, the question would creep back in.

"You study the universe, but do you study yourself?"

He shook his head, annoyed.

"It's a meaningless question," he told himself again. *"It has no answer because it doesn't need one."*

But no matter how much he tried to convince himself...

He wasn't sure he believed that anymore.

The Thought That Won't Leave

Keshav buried himself in work, trying to drown out the lingering question with equations and data.

The hum of servers, the quiet murmurs of engineers in nearby offices, the faint beeping of telemetry monitors—these were the sounds he was used to. They represented order, precision, and clarity.

But today, for the first time, they felt like background noise.

Because beneath it all, something unsettling was happening in his mind.

A question he had dismissed as meaningless had taken root.

"You study the universe, but do you study yourself?"

It was absurd. It was vague. It was the kind of thing philosophers and mystics pondered over in books he had never bothered to read.

And yet, he couldn't stop thinking about it.

Logic vs. Intuition

He tried to rationalize his discomfort.

"This is just a cognitive distraction," he told himself. *"A temporary glitch in the system."*

His mind had been conditioned for logic, for problem-solving, for external inquiry. He wasn't programmed for self-reflection.

He didn't need to study himself—he needed to study what was real.

Planets. Stars. Matter and motion.

These were tangible things. Things that could be observed, measured, calculated.

But how did one study themselves?

What kind of experiment could be conducted? What data could be collected? What conclusions could be drawn?

His entire framework for understanding the world was built on the external.

And maybe that was why the question bothered him so much.

Because it wasn't asking for an external answer.

It was demanding an internal one.

And he had none.

The Unsettling Simplicity of the Question

Most difficult questions were complex.

That was the nature of science—problems required effort, time, and deep understanding to solve.

But this question?

It was too simple.

It didn't require mathematical models or advanced physics. It didn't require years of research or massive funding. It didn't require telescopes, satellites, or algorithms.

All it required was him.

And for the first time, Keshav realized that the simplest questions are often the hardest to answer.

The Inescapable Echo

By mid-afternoon, he was growing frustrated.

He had tried to push it away, to bury himself in his work, but the question kept returning—like an equation he had overlooked, like an anomaly in the data that refused to be ignored.

He went about his usual meetings, reviewing research proposals, discussing upcoming experiments.

But even as his colleagues spoke, he found his mind drifting.

"Do you study yourself?"

It came in quiet moments—in the pauses between conversations, in the brief silences during presentations, in the seconds between reading one report and the next.

No matter how much he tried to suppress it, it was there.

Like a song stuck in his head. Like an unsolved problem taunting him. Like a door that had been opened just a crack—and now refused to shut.

The Resistance to Introspection

He had spent years dismissing self-exploration as unscientific.

He had always believed that people who spent time contemplating themselves were merely indulging in self-importance.

"People spend their entire lives trying to 'find themselves' while the real mysteries are out there—beyond our atmosphere, beyond our galaxy, beyond the limits of our current understanding."

Self-reflection was for those who had no greater purpose.

It was for those who didn't know how to think scientifically.

It was for those who couldn't handle the complexity of reality, so they retreated into their own minds.

Or so he had believed.

But now... he wasn't sure anymore.

Because if self-exploration was meaningless, why was this question affecting him so much?

If it didn't matter, why couldn't he shake it off?

And if he was so sure that he had the right answers about how to live—why was he suddenly questioning everything?

A Quiet Doubt Creeping In

By the evening, Keshav was exhausted—not physically, but mentally.

He had spent the entire day in battle with himself.

It wasn't that he wanted an answer to the question.

It was that he wanted it to disappear.

He wanted to move on, to return to the certainty he had always lived with.

But certainty had a strange quality to it.

Once a crack forms, it spreads.

And the more he tried to ignore it, the more he realized—he had never truly questioned himself before.

The Moment of Realization

As he packed up for the night, he caught his reflection in the darkened glass of his office window.

For a moment, he just stared.

He had spent years looking outward—toward the stars, toward equations, toward the future of human progress.

But had he ever stopped to truly look at himself?

The question wasn't just about introspection.

It was about awareness.

About whether he had ever taken the time to understand his own mind, his own beliefs, his own motivations.

And for the first time, he realized:

He hadn't.

He had spent years dissecting the universe, searching for its secrets.

But he had never once searched for his own.

And that realization terrified him.

Because what if he didn't like what he found?

The Shift from External to Internal

The universe had always been Keshav's obsession.

As a child, he had spent countless nights staring up at the sky, fascinated by the vastness, the endless expanse of stars and galaxies stretching beyond human comprehension.

By the time he was a teenager, he had memorized the distances between planets, the lifecycles of stars, the mathematics of black holes. While others in his class struggled with calculus, he had already moved on to theoretical physics.

He had always looked outward.

The mysteries of the cosmos had been his purpose, his drive, his escape.

And yet, tonight—for the first time in his life—he was looking inward.

And it unsettled him.

What Does It Mean to Study Myself?

Keshav sat in his apartment, staring at his untouched dinner. His thoughts were still tangled in the question he had been avoiding all day.

"You study the universe, but do you study yourself?"

What did that even mean?

He had always assumed that understanding the universe would eventually lead to understanding everything.

After all, wasn't the human mind just another complex system—neurons firing, chemicals reacting, patterns forming?

He had spent years studying celestial mechanics, gravitational forces, quantum fluctuations.

But he had never once stopped to ask, what was driving him?

Why was he so obsessed with answers? Why did he crave knowledge the way others craved wealth or fame? What was he really searching for?

And the most unsettling question of all—was it possible that some truths couldn't be measured?

The Limits of Science

Science had always been his guide, his foundation.

The idea that anything existed outside the realm of scientific inquiry had always seemed ridiculous to him.

The universe operated by laws, whether humans understood them or not.

Everything could be quantified, explained, reduced to logic and numbers.

That was the beauty of science—it removed uncertainty.

But now, for the first time, he found himself facing a question science couldn't answer.

"What does it mean to study myself?"

There was no formula for that. No experiment to conduct. No way to measure the depth of one's own mind.

And that disturbed him more than he was willing to admit.

The First Crack in Certainty

Keshav had always been certain about his beliefs.

He had never doubted that science was the ultimate path to truth.

He had never questioned whether logic alone was enough.

But now, a small crack had formed in that certainty.

A whisper of doubt. A feeling that something had been overlooked.

For years, he had assumed that the greatest questions in existence were the ones written in the stars.

But what if the greatest question was inside him?

A Restless Mind

He paced his apartment, his thoughts spiraling in a way they never had before.

He tried to reason his way out of it.

"I'm just tired. That's all this is."

"I've been overworked. I need a break."

"This is a meaningless distraction from real scientific progress."

But the arguments felt weak.

Because deep down, he knew that this wasn't just a passing thought.

This was something fundamental.

Something that had been waiting for him to notice.

And now that he had, he couldn't unsee it.

A New Kind of Unknown

Keshav had spent his life chasing the unknown.

He had been drawn to the mysteries of dark matter, the unanswered questions of the universe, the places where human understanding had not yet reached.

But now, he was beginning to realize—

There was another kind of unknown.

Not in space. Not in physics. Not in equations.

But in himself.

He had spent years mapping the cosmos.

But had he ever truly mapped his own mind?

Had he ever asked why he believed what he believed?

Had he ever questioned his own perception of reality?

Had he ever tried to understand the forces that shaped his thoughts, his emotions, his choices?

The answer was clear.

No.

The Fear of Looking Inward

The idea of self-exploration unnerved him.

Not because he thought it was pointless.

But because he was afraid of what he might find.

What if his pursuit of knowledge had never been about understanding the universe—but about avoiding something within himself?

What if his obsession with science had been a way to distract himself from deeper questions he had never wanted to face?

What if the real unknown wasn't out there... but inside him?

The thought sent a chill through him.

Because if that was true, it meant something even more unsettling—

He had been running all along.

A Mind at War with Itself

Keshav hated uncertainty.

It was why he had become a scientist—to remove doubt, to answer questions, to bring clarity to the chaos of existence.

But now, he was realizing something terrifying.

Some questions weren't meant to be answered.

Some questions weren't about logic or science or proof.

Some questions simply... were.

And for the first time in his life, he didn't know how to deal with that.

The Shift Had Begun

As he lay in bed that night, staring at the ceiling, he knew something had changed.

He could pretend all he wanted.

He could tell himself this was temporary, meaningless, a distraction.

But he knew the truth.

A shift had begun.

And no matter how much he resisted...

There was no turning back.

The Beginning of Introspection

Keshav lay in bed, staring at the ceiling, his mind refusing to rest.

The room was dark, but his thoughts were bright, relentless, intrusive.

"Do I even know myself?"

The question came uninvited, slipping through the cracks of his rational mind, taking root in places he had never explored.

It was simple. Too simple.

Yet, it carried the weight of something immense.

A truth he had never considered.

A truth he wasn't sure he was ready to face.

A Stranger in His Own Mind

He had always believed he understood himself.

He knew what he wanted, where he was going, what drove him.

His identity had been built on certainty, on logic, on an unshakable foundation of knowledge and ambition.

And yet, as he lay there, he felt like a stranger to himself.

It was a feeling he had never experienced before—a sudden disconnection, as if he was looking at his own life from the outside.

For years, he had studied black holes, distant galaxies, the mysteries of the cosmos.

But now, he felt lost within his own mind.

As if he had spent his whole life looking outward to avoid looking in.

And that realization terrified him.

The Search That Never Included Himself

He had always been a seeker.

From a young age, he had been drawn to the great unknown—searching for answers in the laws of physics, in the fabric of the universe.

But never once had he searched within.

Never once had he asked himself:

"Why do I think the way I do?" "Why do I believe what I believe?" "Who am I beneath all my knowledge?"

He had assumed that self-understanding was a given, that knowing the world meant knowing himself.

But now, he realized how wrong he had been.

Knowing about something was not the same as knowing it.

And knowing the universe was not the same as knowing himself.

Pushing the Thought Away

He turned onto his side, squeezing his eyes shut, willing the thoughts to disappear.

"This is pointless," he told himself. *"This is just exhaustion. Overthinking. Nonsense."*

He had never believed in introspection.

He had always dismissed self-inquiry as philosophical indulgence, a distraction from real knowledge.

But the harder he tried to push the thought away, the stronger it became.

Because something had already changed.

A crack had formed.

A shift had begun.

And he knew—there was no undoing it.

The Weight of Realization

He had spent years defining his identity through external achievements.

But now, he wasn't sure if those achievements meant anything at all.

Because if he didn't know himself...

Then who had been chasing all of this success?

Who had been making these choices?

Who was he, really?

The thought was dizzying.

Because if he had never questioned it before, then maybe—just maybe—he had never truly known.

And that was a terrifying possibility.

The Dawn of a New Journey

The stars had always guided him.

But tonight, he wasn't looking at the stars.

He was looking at something deeper, something closer, something infinitely more complex than the cosmos itself.

His own mind.

For the first time, he wasn't just seeking knowledge.

He was seeking himself.

And this—this was a journey he had never prepared for.

Part II: The Struggle

Between Science, Philosophy, and Meaning

"You study the universe, but do you study yourself?"

4: The Late-Night Debate

Science vs. Meaning

The city was quieter at night.

Keshav sat alone at a café near his apartment, stirring his untouched coffee, his mind still tangled in the thoughts that had followed him home.

It had been three days since the janitor had asked him that question.

"You study the universe, but do you study yourself?"

Three days, and still, he couldn't shake it.

He had tried to lose himself in work, in equations, in the cold certainty of numbers. But no matter how hard he tried, something had shifted.

And now, as he sat in the dimly lit café, watching the late-night crowd filter in and out, he felt like a man standing at the edge of something vast—something he had never dared to explore.

He was so lost in thought that he didn't notice the man approaching his table until a familiar voice cut through the silence.

"Keshav Raghunathan, lost in thought? Now that's something I never expected to see."

Keshav looked up.

Standing before him was **Amit Sharma**—an old friend from his IIT days, and the last person he needed to see right now.

Because if there was one person who thrived on questioning everything, it was Amit.

And tonight, Keshav wasn't sure if he had the answers.

An Unexpected Meeting

Keshav hadn't seen Amit Sharma in years.

Yet, here he was, standing across the café table, his signature half-smile and sharp eyes unchanged.

Amit had always been a philosopher at heart, a man who questioned everything, challenged conventional thinking, and thrived in conversations that others found uncomfortable.

Back in their IIT days, Amit had been one of the few people who could hold his own against Keshav in debates. While Keshav had been immersed in science, logic, and empirical knowledge, Amit had been drawn to philosophy, human consciousness, and the search for meaning.

Theirs had been a friendship built on intellectual clashes—one that had always walked the fine line between respect and irritation.

"Well, are you going to say something, or have you finally discovered that silence is more powerful than words?" Amit said, dropping into the chair across from Keshav.

Keshav sighed. *"Still dramatic, I see."*

"Still avoiding the real questions, I see."

Amit flagged down a waiter, ordering a black coffee as if he had all the time in the world.

"So," Amit continued, leaning back in his chair, *"what's got the great scientist looking like he just stumbled into an existential crisis?"*

Keshav smirked, shaking his head. *"I'm fine, Amit. Just caught up in work."*

"Work, huh?" Amit eyed him carefully. *"And yet, your coffee is cold, your shoulders are tense, and you didn't even look up when I walked in. That doesn't sound like the Keshav I know."*

Keshav exhaled. He should have known Amit would see through him. He always did.

"Alright, fine," Keshav admitted. *"It's just been... a strange few days."*

"Strange? Now you have my attention," Amit said, sitting forward. *"Go on."*

A Familiar Tension

Keshav hesitated.

Did he really want to open this door?

He and Amit had always seen the world differently.

For Keshav, truth was something to be measured, quantified, proven. For Amit, truth was something to be experienced, felt, understood beyond logic.

They had debated this endlessly in college, neither willing to concede ground.

And now, after years of certainty, Keshav wasn't sure he wanted Amit to see even a sliver of doubt creeping into his thoughts.

"It's nothing, really," he said, waving it off. *"Just... a question someone asked me. A simple one, but it stuck with me."*

"Ah," Amit smiled knowingly. *"The most dangerous kind of question."*

Keshav chuckled. *"It was a janitor at my lab, of all people. He asked me—'You study the universe, but do you study yourself?'"*

Amit's eyes gleamed with interest.

"And what did you say?"

"I laughed, of course," Keshav replied. *"It's a meaningless question. What does it even mean to 'study myself'? I study what's real. The universe. The forces that govern existence."*

Amit nodded slowly, as if considering Keshav's words.

Then he smiled, leaned back, and said something that made Keshav's jaw tighten.

"That's the dumbest thing I've ever heard you say."

The Debate Begins

"Excuse me?" Keshav raised an eyebrow.

"Oh, don't get me wrong," Amit said, taking a sip of his coffee. *"It's not dumb because you don't know the answer. It's dumb because you think the question doesn't matter."*

Keshav crossed his arms. *"It doesn't. I study reality. What does 'studying myself' even mean? Consciousness is just a result of biochemical processes in the brain. There's nothing deeper to understand."*

Amit shook his head, amused. *"You see, this is what I love about you, Keshav. You have a brilliant mind, but sometimes you're as narrow as the equations you worship."*

"And you," Keshav shot back, *"are too obsessed with abstract nonsense. The world moves forward because of science, not philosophy."*

"Ah, yes, the classic argument," Amit said, grinning. *"But let me ask you this—if science has all the answers, why do the smartest people in the world still feel lost?"*

Keshav's smirk faded.

"What do you mean?"

"You just said you study what's 'real.' But tell me, Keshav—why do brilliant minds, people who have unlocked the secrets of the universe, still struggle with meaning, purpose, and fulfillment?"

Keshav frowned.

"Because meaning is a human construct," he said. *"It's an illusion. We assign meaning to things because we need to feel in control."*

Amit shook his head.

"And yet, you're sitting here, staring into space, questioning yourself in a way you never have before. Why? If meaning is an illusion, why do you care?"

Keshav had no immediate response.

Because Amit was right.

He did care.

He had been bothered by this question in a way he couldn't explain.

And the fact that he had no logical reason for it unsettled him.

The Argument Deepens

"Science explains everything, Amit," Keshav said, pushing back. *"It may not have all the answers yet, but it will. Consciousness, emotions, purpose—it's all just chemistry and neural activity. It's not some great mystery."*

Amit sighed.

"Tell me something, Keshav. Have you ever been in love?"

Keshav blinked. *"What does that have to do with anything?"*

"Just answer the question."

Keshav hesitated. *"No, not really. I've had relationships, but nothing serious."*

"Okay," Amit said, leaning in. *"Explain love to me. Scientifically."*

"Easy," Keshav said. *"Love is just a series of biochemical reactions—dopamine, oxytocin, serotonin. It's an evolutionary response designed to ensure survival and reproduction."*

Amit smiled.

"And yet, no poet has ever written about dopamine. No artist has ever painted serotonin. No person has ever felt a rush of oxytocin and mistaken it for love."

Keshav exhaled sharply.

"What's your point?"

"My point, dear scientist," Amit said, *"is that just because something can be broken down into biological components doesn't mean its experience isn't real. You dismiss meaning as an illusion. But meaning is as real as love, as grief, as joy. You can measure the brain activity behind these things, but that doesn't change how they feel. And that feeling? That's what makes them real."*

Keshav was silent.

Because he had never thought about it that way.

He had always assumed that if something couldn't be quantified, it wasn't real.

But Amit had just challenged that assumption in a way he couldn't ignore.

The First Moment of Doubt

For the first time in his life, Keshav didn't know how to argue back.

He had spent years dismissing meaning, purpose, and introspection as philosophical distractions.

But now, sitting across from Amit, he felt something he hadn't felt in a long time.

Doubt.

And it terrified him.

The Clash of Perspectives

The café had quieted, the late-night crowd thinning as the hours crept forward. The soft hum of distant conversations and the occasional clatter of dishes filled the air, but at Keshav and Amit's table, a different kind of tension was brewing.

Keshav leaned forward, his fingers wrapped around his coffee cup, the warmth grounding him as his mind prepared to dismantle Amit's argument.

"Look, Amit, science explains how things work because that's the only thing that matters. The 'why' is irrelevant."

Amit smirked, stirring his coffee slowly. *"You say that, but if 'why' doesn't matter, why are we even having this conversation? Why does the smartest mind I know look like he's searching for something?"*

Keshav exhaled sharply. *"People search for meaning because it's a survival mechanism. It's biological, evolutionary. Our brains are wired to seek patterns, to create purpose where none exists. That doesn't mean meaning is real—it just means we need it to function."*

Amit nodded as if he had expected this response. *"Ah, the good old 'meaning is an illusion' argument. But tell me, Keshav—if meaning is an illusion, why do you care about the work you do? Why dedicate your life to space, to discovery, to knowledge?"*

"Because knowledge is real," Keshav shot back. *"Because the universe is real. Because understanding the cosmos helps us advance as a species. Science gives us power, control, progress."*

Amit leaned in. *"Control? Progress? Advancement? You just used words that imply direction, purpose, and value. Tell me, Keshav—if there's no inherent meaning, why do you care about these things?"*

Keshav opened his mouth to respond but hesitated.

Because for the first time, he wasn't sure how to answer.

The Core of the Debate

Keshav took a slow sip of his coffee, buying time. He needed to regain control of the conversation.

"Look, Amit," he said, carefully choosing his words. *"Science has always been about cutting through illusions. It has shown us that the Earth isn't the center of the universe, that time isn't absolute, that life is the result of natural selection, not divine intervention. Science has dismantled every false narrative humans have ever created. Meaning is just another illusion waiting to be exposed."*

Amit nodded, considering his words. *"I agree—science has done incredible things. It has changed the way we see reality. But tell me, Keshav—when Newton discovered gravity, did the force suddenly come into existence, or was it always there?"*

"It was always there," Keshav replied, slightly annoyed by the obviousness of the question.

Amit smiled. *"Exactly. Newton didn't create gravity; he simply uncovered what was already true. So what if meaning is the same? What if it's not an illusion, but something real—something science just hasn't figured out yet?"*

Keshav frowned. *"That's a stretch. You're comparing a physical force with something completely subjective."*

"Am I?" Amit raised an eyebrow. *"Tell me, Keshav—does consciousness exist?"*

"Of course," Keshav replied. *"But we know it's a product of the brain, of neurons and electrical signals."*

"Yes, but does that make the experience of it any less real?"

Keshav hesitated. *"No... but—"*

"Then why are you so sure meaning isn't real, just because you can't measure it yet?"

The question hit like a precision strike to Keshav's logic.

Science and the Search for Meaning

Keshav exhaled, shaking his head. *"Amit, science explains everything. Even if we don't have all the answers yet, we will. There's nothing mystical about meaning—people just need it because they're afraid of chaos."*

Amit smirked. *"You keep saying science explains everything, but does it? If it did, why are you still searching?"*

Keshav clenched his jaw. *"I'm not searching. I'm just—"*

"Unsettled? Conflicted?" Amit interrupted. *"Tell me, Keshav—if science explains everything, why do the most intelligent people in the world still struggle with purpose? Why do the greatest minds—scientists, philosophers, thinkers—spend their lives wrestling with meaning?"*

"Because they haven't fully accepted reality," Keshav said, frustrated now. *"Because people don't want to accept that we're just chemical reactions and electrical impulses."*

Amit nodded. *"And you? Have you accepted it?"*

Keshav stared at him. The answer should have been yes.

But the fact that he hesitated...

The fact that he had spent days wrestling with a simple question from a janitor...

The fact that he was sitting here, debating Amit instead of dismissing him outright...

It meant something.

And he didn't like that.

The Moment of Doubt

Amit leaned back, studying him.

"You know what I think, Keshav?" he said finally. *"I think deep down, you know science isn't enough. You don't want to admit it because it would mean questioning everything you've built your life on. But something is changing in you, whether you like it or not."*

Keshav scoffed, but the reaction felt forced. *"Don't psychoanalyze me, Amit."*

"I don't need to," Amit said, smiling. *"I can see it. You're not just debating me—you're debating yourself."*

Keshav looked away, focusing on the steam rising from his coffee.

"Tell me one thing," Amit said, his voice quieter now.

"If intelligence alone was enough... wouldn't you already have found what you're looking for?"

Keshav froze.

Because that was the question that had been lurking in the back of his mind.

And the fact that he didn't have an answer for it...

That was the most unsettling realization of all.

A Question That Won't Leave

The debate had ended, but Amit had planted a seed.

As Keshav walked home that night, his thoughts were louder than ever.

For years, he had believed that science was enough—that the only truth that mattered was the kind that could be measured, observed, and quantified.

But now, he wasn't sure.

Because despite having all the knowledge he had ever wanted...

There was still something missing.

And for the first time in his life, he was beginning to think that maybe—just maybe—Amit was right.

Maybe the greatest unknown wasn't in the universe.

Maybe it was within himself.

And that thought terrified him.

Science vs. Meaning – The Core Argument

The night air outside the café was crisp, but inside, the tension between Keshav and Amit had turned electric.

Their debate had moved past casual disagreement—it was now a full-blown ideological clash.

Keshav sat back, exhaling sharply, frustration flickering in his eyes. He had spent years perfecting his ability to argue, to use logic as an unshakable weapon. Yet, tonight, Amit was proving to be an opponent unlike any other.

"Truth must be measurable, Amit," Keshav said, his voice firm. *"Anything that cannot be quantified is subjective. That's the entire foundation of science—objectivity. Measurement. Proof."*

Amit smirked, unfazed. *"And yet, the things people value most in life—love, beauty, meaning—none of them can be measured. Does that make them any less real?"*

Keshav frowned. *"That's different."*

"Is it?" Amit leaned in, his gaze sharp. *"Tell me, Keshav—if truth is only what can be measured, what is the value of a human life?"*

Keshav crossed his arms. *"That's an emotional argument. The value of life is defined by biology—our ability to survive and reproduce, to ensure the continuation of the species. Everything beyond that is just an artificial construct."*

"Ah," Amit nodded slowly. *"So you're saying that if something isn't necessary for biological survival, it has no real value?"*

"Exactly."

Amit let out a short laugh, shaking his head. *"Then explain art. Music. Poetry. Why do humans create things that serve no evolutionary function?"*

Keshav hesitated. *"They do serve a function. Art and music are forms of communication, of social bonding. They strengthen group cohesion, which indirectly benefits survival."*

Amit smiled. *"So you're saying that a symphony, a timeless novel, or a masterpiece painting is just an advanced form of tribal bonding?"*

"Essentially, yes," Keshav said. *"Humans assign unnecessary meaning to things because it makes life easier to navigate. But meaning itself is not real."*

Amit sighed, setting his coffee cup down carefully. *"You really believe that?"*

"Yes."

"Then why do you look so unconvinced?"

The Crisis of Intelligence

Keshav's jaw tightened. He didn't like this. Amit had a way of digging into the cracks of his logic, forcing him to confront ideas he didn't want to consider.

"Listen, Amit," Keshav said, leaning forward. *"Intelligence is about understanding reality as it is, not as we want it to be. If meaning were real, science would have discovered it by now."*

Amit didn't respond right away. Instead, he studied Keshav, as if searching for something deeper beneath the words.

Then, he asked the question that changed everything.

"Then why does intelligence, which should bring clarity, often lead to existential crises?"

Keshav blinked. *"What?"*

"If intelligence is the highest form of understanding, why does it so often lead to doubt, anxiety, and meaninglessness?" Amit asked. *"Why do the most brilliant minds—scientists, philosophers, thinkers—struggle with depression, loneliness, and emptiness? If knowledge is the answer, why do the people who know the most suffer the most?"*

Keshav didn't respond.

Because he knew Amit was right.

He had seen it firsthand.

Brilliant scientists who had unlocked the secrets of the universe—but couldn't find happiness in their own lives. Engineers who had built machines that defied human limits—but struggled to find purpose in their own existence. Intellectuals who had mapped the cosmos—but felt lost in their own minds.

He had always assumed that this was a side effect of high intelligence—an unfortunate trade-off for brilliance.

But what if it was something else?

What if knowledge wasn't enough?

What if meaning wasn't an illusion—but an essential part of being human?

Beyond Logic

"Keshav, you've spent your life looking outward," Amit continued, his voice softer now. *"You've searched for truth in equations, in space, in the mechanics of the universe. But have you ever searched for truth within yourself?"*

Keshav clenched his fists. *"That's philosophy, Amit. It's not real."*

"Is it not?" Amit tilted his head. *"What makes something real, Keshav? Can you measure love? Can you quantify suffering? Can you define consciousness in a way that fully explains the experience of being alive?"*

Keshav exhaled through his nose, frustration bubbling.

"We may not have all the answers yet, but we will. Science is progressing. Eventually, we'll understand everything."

"Maybe," Amit said. *"But tell me this—if one day science could explain every function of your brain, every emotion, every thought down to its most basic chemical reaction... would that make your experience of life any less real?"*

Keshav hesitated.

Because no. It wouldn't.

Even if he knew exactly how dopamine caused happiness, it wouldn't make the feeling any less alive when he experienced it.

Even if he understood the precise neural pathways that processed grief, it wouldn't make loss any less painful.

Meaning wasn't something he could measure.

But that didn't mean it wasn't real.

The Fundamental Human Experience

Amit leaned forward. *"Keshav, you think meaning is just a human illusion. But what if it's the opposite? What if meaning is the only thing that makes being human bearable?"*

Keshav looked away.

Because for the first time, he wasn't sure anymore.

Amit continued. *"You say that intelligence is the path to truth. But the more you know, the less certain you seem. Why?"*

Keshav ran a hand through his hair. He had no response.

Because he had spent the last few weeks feeling exactly that.

The more he understood, the less complete he felt.

The more knowledge he accumulated, the further away he seemed from any real sense of fulfillment.

It was a paradox.

And it was shaking the foundation of everything he had believed.

A Crack Too Deep to Ignore

The conversation had ended.

But Amit had left Keshav with something he couldn't walk away from.

The realization that he didn't have an answer.

That science had not given him the clarity he once thought it would.

That meaning—something he had dismissed his entire life—might be more real than he ever believed.

As he walked home that night, his mind was no longer just questioning.

It was searching.

And he had no idea where that search would take him.

Keshav's First Moment of Doubt

The walk home was longer than it should have been.

Keshav moved through the quiet streets, his footsteps slow, his mind heavier than it had ever been. The conversation with Amit replayed over and over, each word lingering like an unresolved equation, each argument breaking through the walls he had spent years constructing.

For the first time in his life, logic felt insufficient.

That realization disturbed him more than anything.

The Collapse of Certainty

Keshav had built his life on certainty.

Science had always provided a clear structure, a system that worked.

Every problem had a solution. Every unknown had an answer, waiting to be uncovered. Every mystery could be unraveled through study, observation, and logic.

Yet, tonight, something had changed.

Tonight, he had encountered a question that logic could not answer.

"If intelligence is the path to truth, why does it so often lead to doubt?"

"If knowledge is enough, why do the most brilliant minds still feel lost?"

Amit's words had cut deeper than he wanted to admit.

Because they were not just philosophical musings.

They were questions he had been avoiding asking himself.

And now that they had surfaced, he could not escape them.

The Unraveling of His Belief System

He reached his apartment and collapsed into his chair, staring blankly at his laptop screen.

He wanted to escape this feeling.

He wanted to lose himself in equations, in research, in the comfort of data and numbers.

But for the first time, they didn't feel comforting.

For the first time, they felt... irrelevant.

The formulas, the simulations, the research reports—none of them could answer the question growing inside him.

Because this question wasn't external.

It wasn't about the laws of physics, the movement of planets, or the nature of space-time.

It was about him.

And he had no equation for himself.

The Illusion of Control

Keshav had always believed that intelligence gave him control.

Control over his work. Control over his life. Control over the way he understood the world.

But now, he felt something he had never felt before.

Powerless.

Because the tools he had always relied on—logic, reasoning, science—felt inadequate for the first time.

He tried to fight it.

He reminded himself that science had always found answers.

That every great mystery of the past—from the movement of the planets to the nature of light—had once been unknown but was now understood.

But what if this was different?

What if the search for meaning wasn't just another problem to be solved?

What if meaning wasn't a scientific equation but a human experience?

And what if—despite everything he had achieved, everything he had learned—he had missed the most important thing of all?

Knowledge vs. Understanding

For years, he had worshipped knowledge.

He had believed that the more he knew, the more complete he would become.

That if he simply gathered enough information, all the pieces of the puzzle would fall into place.

But now, he was beginning to see the flaw in that thinking.

Because knowledge was not the same as understanding.

A scientist could study emotions, but that didn't mean he understood what it felt like to grieve. A physicist could explain the mechanics of a sunset, but that didn't mean he had ever truly admired one. A doctor could understand the chemistry of love, but that didn't mean he had ever been in love.

Science could describe reality, but could it experience it?

And if not, was there something beyond science?

A Crack Too Deep to Ignore

He sat in silence for what felt like hours.

At some point, he realized he didn't want to fight it anymore.

The doubt was there.

It had taken root, whether he wanted it to or not.

And trying to force it away wouldn't change anything.

He had spent years dismissing philosophy, meaning, and introspection as illusions.

But what if they weren't illusions?

What if they were a different kind of truth—one that science could never fully explain?

The Fear of Letting Go

A part of him wanted to resist.

Wanted to hold on to the certainty of logic, to the structure of science, to the belief that everything could be explained.

Because if he let go of that certainty... what was left?

He didn't know.

And that terrified him.

Because for the first time in his life, he was no longer sure that knowledge was enough.

The Search Begins

Keshav leaned back, closing his eyes.

A week ago, he would have dismissed all of this as nonsense.

But now, he couldn't.

Because the cracks in his belief system were no longer small.

They were deepening, widening, breaking apart the foundation he had always stood on.

And there was no going back.

He didn't have answers yet.

But for the first time, he was starting to think that maybe—just maybe—the greatest search of all was not for knowledge of the universe.

It was for knowledge of himself.

And that was a search he had never prepared for.

A Question That Stays with Him

The café had emptied out, leaving only the soft hum of jazz playing in the background and the distant clatter of dishes from the kitchen.

Keshav sat across from Amit, his fingers lightly tapping against the table, his mind running in circles. The conversation had gone further than he expected, pushing into territory he had never explored, never wanted to explore.

Then came Amit's final question.

"If intelligence alone was enough, wouldn't you already have found fulfillment?"

Keshav felt his breath catch slightly.

Because this time, he didn't have an answer.

The Silence That Followed

Amit watched him carefully, his expression calm, almost expectant. He wasn't pressing for an immediate response.

And Keshav hated that.

Because if Amit had kept arguing, he could have countered. He could have pulled out another scientific explanation, another logical framework, another way to dismiss the importance of meaning.

But Amit had stopped.

Because he knew—the question itself was enough.

Keshav opened his mouth, then closed it. He wanted to say something—anything.

To deny, refute, dismantle the idea that intelligence alone wasn't enough.

To say that he was fulfilled, that knowledge had always been enough.

But the truth was he couldn't say it.

Not because he wasn't sure.

But because, deep down, he knew he would be lying.

A Realization He Wasn't Ready For

For years, Keshav had believed that intelligence was the highest form of power.

That knowledge, understanding, and logic were the ultimate tools of human evolution.

That once you had mastered science, reason, and intellect, there was nothing more to seek.

But now, sitting across from Amit, he realized—he had mastered all those things.

And yet, he was still searching.

Why?

If knowledge alone brought clarity, why was he questioning everything now? If science was the ultimate truth, why did it feel incomplete? If intelligence was the key to fulfillment, why hadn't he already found it?

It was a thought he had never dared to entertain before.

But now, he couldn't escape it.

The End of the Debate, But Not the End of the Question

Amit took one last sip of his coffee, then set the cup down gently.

"You don't have to answer now, Keshav," he said, his voice quieter now. *"But one day, you will have to."*

Keshav let out a slow breath, forcing himself to meet Amit's gaze. *"I'll think about it,"* he admitted.

"Good," Amit said, pushing back his chair. He stood up, stretching slightly. *"That's all I wanted."*

Keshav watched as his friend pulled on his jacket, paid for his coffee, and walked toward the door.

Just before stepping outside, Amit turned back.

"You know," he said with a small smile, *"not all truths can be calculated. Some have to be realized."*

Then, he was gone.

And Keshav was left alone with his thoughts.

The Walk Home – A Mind in Turmoil

Keshav stepped out into the cold night air. The streets were nearly empty now, the city settling into silence.

He walked slowly, his hands in his pockets, his mind far from settled.

Amit had left him with something far more dangerous than an argument.

A doubt.

One that wouldn't go away.

Because doubt wasn't just a question. It was a crack. And cracks had a way of spreading.

A Question He Can't Ignore

Keshav had always believed that knowledge was enough.

That if he just kept seeking, kept learning, kept expanding his understanding of the universe, he would eventually find clarity.

But tonight, he realized something he had never considered before.

What if clarity wasn't the same as fulfillment? What if intelligence wasn't the same as wisdom? What if knowledge wasn't enough?

He had spent his life searching outward.

For the first time, he wondered if the real search had to begin inward.

And that thought...

Terrified him.

The Beginning of Something New

As he reached his apartment, he stood at the door for a moment before stepping inside.

The silence in his home felt different now.

It wasn't just the absence of noise.

It was the presence of something else.

Something unspoken. Something waiting to be understood.

He set down his bag, rubbing his temples.

He had entered the night as a man of certainty.

He was leaving it as a man of questions.

And for the first time in his life, he didn't know if that was a step forward or backward.

But one thing was clear—his search was no longer just about the universe.

Now, it was about him.
And that...
Was a journey he had never prepared for.

5: The Illusion of Free Will

The question had stayed with him.

It had followed him home, settled in the corners of his mind, and refused to leave.

"If intelligence alone was enough, wouldn't you already have found fulfillment?"

Even now, as Keshav sat in his office the next morning, staring at the complex models running on his screen, the doubt lingered.

He had spent his life believing in certainty, in logic, in the power of the human mind to uncover every truth.

But if intelligence could be dissected, analyzed, predicted—was it really free?

His own thoughts felt less like conscious choices and more like inevitable conclusions of his experiences, his knowledge, and his past.

And if that was true...

Did he even have free will?

A knock on his office door broke his thoughts.

"Keshav, you have a meeting with Dr. Malhotra in ten minutes."

Keshav nodded absentmindedly. He had almost forgotten.

Dr. Rohan Malhotra.

An expert in Artificial Intelligence and neuroscience, specializing in the cognitive patterns of decision-making.

Someone who had spent years trying to prove that humans were just biological machines—nothing more than complex programs running on neural circuits.

Keshav closed his laptop and stood up.

He had always dismissed those ideas.

But after last night's conversation with Amit, he wasn't so sure anymore.

And today, he would have to face the question head-on.

The Question of Free Will

Keshav sat at his desk, but his mind wasn't in the room.

His fingers hovered over his keyboard, a half-written email glowing on his screen, waiting to be completed. His usual efficiency was missing. Instead, his thoughts were elsewhere—deep in the uncertainty that had begun to unravel his belief system.

Amit's question had left a mark.

"If intelligence alone was enough, wouldn't you already have found fulfillment?"

It was supposed to be an easy question to dismiss.

But it wasn't.

Now, for the first time in his life, Keshav wasn't just questioning meaning—he was questioning himself.

If science and logic were enough, then why was his mind still restless?

If knowledge was the highest pursuit, why did he feel like he had reached the peak only to find an empty summit?

He leaned back in his chair, rubbing his temples.

Maybe he was asking the wrong question.

Maybe this wasn't about meaning at all.

Maybe this was about control.

Choice.

Free will.

The Illusion of Choice

For years, Keshav had believed that every decision he made was his own.

From choosing his career in space research to every equation he had solved—his intellect, his reasoning, his own mind had guided him.

But what if that wasn't true?

What if everything he had done, every conclusion he had reached, every step he had taken in life... had already been decided before he even made the choice?

The thought sent a ripple of discomfort through him.

He had always believed in logic, in autonomy, in the human ability to shape one's own path.

But now, he wasn't so sure.

He turned his chair slightly, looking at the books on his shelf—texts on physics, mathematics, neuroscience, artificial intelligence.

He had spent years studying how things worked.

But had he ever questioned why they worked that way?

Had he ever stopped to wonder if he was simply a product of his biology, his environment, and his past?

And if that was true...

Was he really free?

Are We Just Biological Machines?

He had spent his career working with patterns, predictions, and probabilities.

He had seen how behavioral algorithms could predict human decisions with astonishing accuracy.

Wasn't that proof that free will was an illusion?

After all, science had shown that:

- Neuroscience could predict a decision seconds before a person was even aware they had made it.
- Behavioral studies could determine how people would react in certain environments, proving that external factors shaped "choices."
- Genetics played a role in intelligence, personality, and even emotional responses—preprogramming people to a degree they were unaware of.

So, was he really in control?

Or was he just responding to inputs like an advanced machine?

Was his entire life just the result of physics, biology, and past experiences acting on him?

If that was true...

Then what did it mean to be human?

The Discomfort of Determinism

Keshav shifted uncomfortably in his chair.

He had spent years believing in certainty.

But now, he was beginning to see the flaw in his thinking.

If everything was determined by biology, physics, and external forces, then was anything truly his choice?

Had he ever had control over his own thoughts?

Had he ever been free?

The idea disturbed him.

Because if free will was an illusion, then what was the point of anything?

If he was simply following a pre-written script, then was his entire life nothing more than an inevitable outcome of past events?

And if that was true...

Then what was the purpose of seeking knowledge, of discovery, of progress?

What was the point of ambition?

What was the point of anything?

A Growing Realization

Keshav exhaled slowly, staring at his hands.

His mind was his greatest tool.

It had always been his source of power, of clarity.

But now, for the first time, he felt like a prisoner inside his own thoughts.

If every choice he made was simply the result of external factors, then he was no different than a complex machine—an advanced biological program, following a set of instructions written long before he was born.

The thought shook him.

Because it meant that his beliefs, his desires, his entire identity—were they even his?

Or were they just the inevitable result of everything that had come before him?

His upbringing. His genetics. His experiences.

And if that was true...

Then was his very sense of self just an illusion?

The Debate to Come

A knock on the door pulled him out of his thoughts.

"Dr. Raghunathan?"

He turned. One of the interns stood hesitantly in the doorway.

"Your meeting with Dr. Malhotra is in five minutes."

Keshav blinked, momentarily disoriented.

"Right. Thank you."

As the intern left, Keshav took a deep breath and stood up.

Dr. Rohan Malhotra.

A leading AI and neuroscience researcher—a man who had spent years arguing that humans were nothing more than biological machines.

Today, they were supposed to discuss predictive AI models for space systems.

But now, Keshav realized—he had a much bigger question to ask.

Because if anyone could answer whether free will was real...

It was a man who had spent his life trying to prove that it wasn't.

A Step into the Unknown

As Keshav walked toward the meeting room, his thoughts raced.

For years, he had believed in certainty, logic, and control.

Now, he wasn't sure if he had any of them.

And the worst part?

For the first time, he wasn't sure if he even wanted the answer.

Because if Dr. Malhotra confirmed his worst fear—that free will was nothing more than an illusion...

Then everything Keshav had built his life on might collapse entirely.

And he wasn't sure if he was ready for that.

The AI Researcher's Perspective

Keshav entered the conference room, adjusting his thoughts, trying to compose himself.

He had met Dr. Rohan Malhotra only a handful of times, mostly in professional settings. Malhotra was one of the leading AI researchers in cognitive models, specializing in how machine learning mimicked human decision-making.

Today's meeting was supposed to be about predictive AI models for space exploration, a subject Keshav had been deeply involved in.

But after his conversation with Amit, the meeting had taken on a different weight.

Because if there was anyone who could challenge him further on the concept of free will, it was Dr. Malhotra—a man who had spent his life arguing that humans were nothing more than biological machines.

The Man Who Believed in Determinism

"Dr. Raghunathan, welcome," Malhotra greeted, standing up from his chair as Keshav entered.

He was a tall, thin man with sharp eyes—the kind of gaze that dissected things before they were spoken. His office was filled with books on neuroscience, artificial intelligence, behavioral psychology, and cybernetics.

Keshav nodded, shaking his hand.

"Dr. Malhotra. It's been a while."

"Indeed," Malhotra said, motioning toward the seat across from him. *"Shall we begin?"*

As Keshav sat down, Malhotra launched directly into his topic, speaking with precision.

"We've made significant advancements in our AI cognitive models," he said. *"We're now able to predict human decision-making patterns with over 85% accuracy in controlled simulations."*

Keshav raised an eyebrow. *"85%?"*

"And climbing," Malhotra confirmed. *"We're at the stage where, given enough behavioral data, we can predict with high certainty how an individual will respond to a given scenario. In a way, we are proving that what we think of as 'free will' is, in fact, entirely predictable."*

Keshav folded his arms, already sensing where this was going. *"You're saying that humans are just running a program—just like AI?"*

Malhotra gave a slight smile, as if he had been waiting for this question.

"Not 'just like AI,'" he corrected. *"But fundamentally, yes. Every thought you have, every action you take, is the result of pre-existing neural pathways shaped by genetics, experiences, and biological programming."*

"So, you're arguing that humans don't actually make choices?" Keshav challenged.

"I'm arguing," Malhotra said, *"that what you call 'choice' is merely the illusion of choice. Your brain is constantly computing probabilities, weighing past experiences, biological impulses, and learned behaviors to generate what you call a 'decision.' You don't choose—it is chosen for you by your own neural circuitry."*

Keshav exhaled, already feeling the weight of the debate forming.

Are We Just Advanced Machines?

"That's a bold statement," Keshav said, leaning forward. *"You're essentially saying that everything we do, think, and believe is pre-determined?"*

"Yes," Malhotra replied smoothly. *"We call it 'decision-making,' but in reality, the process is automatic. Just like AI follows a structured algorithm based on data inputs, your brain does the same—except your 'data inputs' are your past experiences, emotions, and subconscious impulses."*

Keshav smirked. *"If that's true, then why do people struggle with decisions? If it's all pre-calculated, shouldn't choices be effortless?"*

Malhotra chuckled. *"Ah, but you're confusing effort with free will. The struggle you feel is simply your brain weighing competing variables before reaching a predetermined outcome. The discomfort of choice isn't freedom—it's just the brain running a more complex computation than usual."*

Keshav frowned, resisting the urge to immediately reject Malhotra's reasoning.

Because deep down, he knew that modern neuroscience supported this idea.

Studies had shown that the brain registers a decision before a person is even consciously aware of making it.

Experiments had demonstrated that when people "chose" an option, brain scans could predict the decision seconds before they became aware of it.

It was disturbing evidence.

Because it suggested that what we call 'choice' is actually just an afterthought—our brain's way of justifying a decision it had already made.

The Human Paradox: Intelligence and Emotions

"Fine," Keshav said, rubbing his chin. *"Let's assume for a moment that we're just biological machines. If intelligence is purely mechanical, then explain this—why do humans experience existential questioning? Why do we feel emotions, doubt, conflict?"*

"Ah," Malhotra nodded as if he had anticipated this. *"That's a common misunderstanding. You assume that emotions are something 'special.' But in reality, emotions are just biochemical signals designed to guide behavior. They are not proof of free will—they are proof of programming."*

Keshav shook his head. *"But AI can be programmed to mimic emotions. That doesn't mean it actually 'feels' them."*

"Yet," Malhotra said pointedly.

"Yet?"

"Who's to say that as AI advances, it won't develop self-awareness? Right now, AI lacks subjective experience because it is limited by its architecture. But if intelligence is just a matter of complex processing, then there is no fundamental reason why an advanced AI couldn't eventually 'experience' something akin to human thought and emotion."

Keshav exhaled, feeling the walls of his reasoning being pushed in uncomfortable directions.

"Even if AI mimics human emotions," he argued, *"it doesn't mean those emotions are real. A machine can calculate probabilities, but it doesn't long for meaning or struggle with its own existence."*

"Are you so sure?" Malhotra countered. *"If a machine one day asks itself, 'Why am I here?'—would you still call that an illusion?"*

Keshav paused.

Because that question...

That question hit closer than he wanted to admit.

The Limits of Science

"Here's the problem, Keshav," Malhotra continued, his tone almost sympathetic. *"You're searching for an answer that science may never provide. Science can tell us how the brain functions, how intelligence develops, how neurons fire. But can it truly explain what it means to experience life?"*

Keshav felt the weight of that statement settle deep in his chest.

He wanted to say yes.

Are We Just Pre-Programmed?

The conversation had taken a sharp turn into unsettling territory.

Keshav had entered this debate with confidence, believing that logic, reason, and science would once again validate his worldview.

But now, as Dr. Malhotra spoke, he wasn't so sure anymore.

Every argument Malhotra made seemed to push deeper into an uncomfortable truth—that free will might not be real.

The Science of Decision-Making

"Let's strip this down to the basics," Malhotra said, leaning forward. *"What do we know about how the brain works?"*

Keshav crossed his arms. *"The brain processes information, responds to stimuli, and generates thoughts based on neural activity."*

"Correct," Malhotra nodded. *"But here's where it gets interesting. Neuroscience has shown that our brains make decisions before we're even aware of them."*

Keshav raised an eyebrow. *"Explain."*

Malhotra pulled out his tablet and swiped to a set of notes.

"Take the famous experiment by Benjamin Libet," he began. *"Participants were asked to move their hands whenever they felt like it, while their brain activity was measured. What the experiment found was shocking—brain scans detected the decision being made nearly half a second before the person was consciously aware of 'choosing' to move their hand."*

Keshav had heard of the study before.

Libet's experiment had long been used to support the argument that free will was an illusion.

"So what does that prove?" Keshav asked.

"It proves that what we think of as 'conscious decision-making' is actually just a delayed awareness of an action that has already been determined by neural processes."

Malhotra sat back, studying Keshav's reaction.

"In other words, by the time you 'think' you're making a choice, your brain has already made it for you."

Keshav exhaled slowly, absorbing the weight of that statement.

Was he really making choices?

Or was he just observing the choices his brain had already made?

Predicting Human Behavior: Are We That Predictable?

"Alright," Keshav said, shaking off the discomfort. *"Even if unconscious processes influence decisions, that doesn't mean everything is predetermined. People still change their minds, weigh options, and make unpredictable choices."*

Malhotra smiled knowingly. *"Do they?"*

He tapped on his tablet, bringing up a set of data charts.

"This is what my team has been working on," he explained. *"We've developed AI models that can predict human choices with incredible accuracy. Based on past behavior, emotional responses, and cognitive tendencies, we can determine what a person is likely to do in a given situation. And guess what? There's nothing random about it."*

Keshav frowned, intrigued but skeptical. *"How accurate are we talking?"*

"Over 90% in controlled settings," Malhotra answered confidently.

"90%?"

"Yes. Give me enough data—your habits, your preferences, your neurological patterns—and I can predict how you'll react in almost any given scenario."

Keshav narrowed his eyes. *"That's impossible. Humans are complex. You can't possibly account for every factor."*

Malhotra chuckled. *"Complex? Yes. Unpredictable? No."*

He continued. *"If I know a person's childhood, their genetic disposition, their environmental influences, and their neural activity, I can map out the trajectory of their life with alarming accuracy. What they like, what they dislike, who they are drawn to, what decisions they will make—it all follows a pattern."*

Keshav shook his head. *"That still doesn't mean free will doesn't exist."*

"Keshav," Malhotra said, tilting his head. *"If I can predict your choices before you make them, how is that free will?"*

Keshav opened his mouth to respond—but no words came.

The Illusion of Choice

Malhotra continued, his voice calm but firm.

"You assume you're making choices freely, but in reality, you are just responding to an intricate web of pre-existing conditions."

He raised his hand, counting off points with his fingers.

"Your genes determine your personality traits and cognitive abilities."

"Your childhood experiences shape your values and fears."

"Your culture and upbringing define your biases and beliefs."

"Your brain chemistry dictates your emotional responses."

He lowered his hand.

"So tell me, Keshav—where exactly is the part where you are making a 'free' choice?"

Keshav felt his pulse quicken.

He wanted to fight back. To deny it. To dismiss it.

But a part of him knew that Malhotra wasn't wrong.

Everything he had ever done, every decision he had ever made—was it all just an inevitable outcome of everything that had come before it?

Was he just following a script written by biology and circumstance?

If so, was he ever really in control?

The Emotional Dilemma

Keshav finally spoke, his voice quieter now.

"Even if all of this is true... how do you explain emotions? How do you explain self-awareness?"

Malhotra nodded, as if expecting this.

"Emotions? They're just neurochemical reactions designed to drive behavior."

"And self-awareness?"

"An advanced function of cognition," Malhotra said simply. *"Nothing more than the brain processing its own information."*

Keshav clenched his jaw.

"But AI can mimic emotions," he argued. *"That doesn't mean it actually feels them. That doesn't mean it has an inner experience."*

"Yet," Malhotra corrected.

"Yet?"

"AI lacks human-like emotions because its neural architecture is limited. But if we ever build a system complex enough, capable of simulating the full range of cognitive processes, who's to say it won't develop consciousness?"

Keshav exhaled sharply.

"You really believe AI can become sentient?"

"I believe that consciousness is just a byproduct of complex information processing."

Keshav shook his head.

"That doesn't explain why humans long for meaning. Why we question our own existence. Why intelligence—rather than making us content—seems to create more confusion, more doubt."

"Because," Malhotra said, *"intelligence is not designed to bring clarity. It's designed to solve problems. And when there are no external problems left, it turns inward—searching for meaning where none exists."*

Keshav froze.

Because that was exactly what had been happening to him.

He had spent his life searching outward, trying to understand the universe.

And now that he had reached a point of knowing so much...

His mind had turned inward, searching for something else.

Something science couldn't give him.

The Debate That Wouldn't Leave Him

Keshav stood up, signaling the end of the conversation.

"This has been... insightful, Dr. Malhotra."

Malhotra smirked. *"Insightful? That's a polite way of saying 'unsettling.'"*

Keshav chuckled, but there was no humor in it. *"Maybe."*

As he stepped out of Malhotra's office, his mind was no longer debating the logic of it all.

He had seen the scientific evidence.

He had heard the rational explanations.

And yet, something still didn't sit right.

Because if everything was just pre-programmed...

If free will was just an illusion...

Then why did he feel like he was standing at a crossroads?

And why did he still feel like he had a choice?

The Emotional Paradox

Keshav left Malhotra's office with a mind that felt heavier than before.

The conversation had taken him deeper into the unsettling possibility that everything he had believed about free will—about human consciousness itself—was an illusion.

But something still didn't sit right.

It wasn't just logic that governed human experience.

It was emotion.

He could accept that thoughts might be pre-programmed, that neural activity might be just a series of electrical impulses.

But if that was true, why did emotions feel so undeniably real?

Why did grief hurt? Why did joy feel like lightness? Why did love create a sense of connection beyond logic?

Science could describe how emotions happened, but could it explain why they felt the way they did?

The Limits of Explanation

That evening, Keshav found himself sitting at his desk, staring at a blank document on his laptop.

He had opened a new file intending to write down his thoughts—to rationalize them, analyze them, break them apart until they made sense.

But nothing came.

The problem wasn't that he didn't have the facts.

The problem was that the facts weren't enough.

Neuroscience could map the exact brain regions responsible for happiness, fear, love, and sorrow.

It could explain how neurotransmitters triggered emotional responses based on external stimuli.

It could even predict, in some cases, how a person might react emotionally in a given situation.

But it still couldn't tell him why love felt like love... or why loss felt like loss.

It could describe the mechanics.

But could it ever capture the experience?

And if it couldn't...

Was something missing?

The AI Debate: Do Machines Feel?

The next morning, Keshav found himself back in Malhotra's office, this time with a single question in mind.

"Dr. Malhotra, we need to talk about emotions."

Malhotra smiled as if he had been expecting this. *"Ah. That didn't take long."*

Keshav ignored the remark. *"I understand how the brain processes emotions. I understand that they're biochemical reactions designed to regulate behavior. But if that's all they are... why do they feel so real?"*

Malhotra leaned back, folding his arms. *"They feel real because that's what evolution has shaped them to do. Feeling is just an illusion created by brain chemistry."*

"But if emotions are just programmed responses," Keshav pushed, *"why can't we replicate them in AI? Why can't machines feel?"*

Malhotra raised an eyebrow. *"And who says they can't?"*

Keshav blinked.

"You're saying AI can feel emotions?"

"I'm saying," Malhotra replied, *"that we may just be unwilling to accept that intelligence and self-awareness are nothing more than processes. If humans are simply biological machines running advanced neural programs, why couldn't AI one day develop emotions as well?"*

Keshav shook his head.

"That's different. AI can simulate emotions—it can recognize patterns, respond appropriately, and even mimic human-like behaviors. But that's not the same as feeling."

"Isn't it?" Malhotra challenged. *"What makes you so sure? How do you know that when an AI model reacts with empathy, it isn't actually experiencing something equivalent to what we call 'emotion'?"*

"Because AI doesn't have subjective experience," Keshav argued. *"It processes data, but it doesn't 'feel' joy or sadness the way we do."*

"How do you know?" Malhotra countered. *"How do you know that what you call 'emotion' isn't just a form of data processing in your own brain?"*

Keshav frowned.

"Because I experience it."

"Exactly," Malhotra said, pointing at him. *"You're relying on subjective experience to define what is real. But that's exactly what a machine might do if its intelligence reaches a certain*

level. You assume AI lacks emotion because it doesn't express it like you do. But tell me, Keshav—if an AI model, over time, began expressing self-awareness, emotional distress, or even existential questions, would you still call it an illusion?"

Keshav sat back.

The question unsettled him.

Because if Malhotra was right...

Then maybe the distinction between humans and machines wasn't as clear as he had always believed.

The Subjective Experience Dilemma

"There's a difference," Keshav insisted. *"AI may appear to feel something, but it doesn't actually experience it. It has no inner world. It doesn't suffer. It doesn't long for meaning. It doesn't question its own existence."*

"Yet," Malhotra corrected.

"Yet?"

"Who's to say AI won't develop those things in the future?" Malhotra said. *"Who's to say that intelligence, when complex enough, doesn't naturally begin to search for meaning?"*

Keshav felt a strange unease settle over him.

Because that was exactly what had been happening to him.

The more he knew, the more he questioned. The more he understood, the less certain he became.

Had his own intelligence led him to the same existential crisis that an advanced AI might one day experience?

And if so...

Did that mean intelligence wasn't designed to bring clarity—but to create confusion?

The Search for Something More

Keshav exhaled, running a hand through his hair.

"You still haven't answered my question, Malhotra. Why do emotions feel real?"

Malhotra sighed, almost sympathetically.

"Because, Keshav... you are designed to believe they are. If you didn't, the system wouldn't work."

Keshav frowned. *"The system?"*

"Human behavior," Malhotra clarified. *"Emotions are just regulatory mechanisms. They exist to drive action. If pain didn't feel real, you wouldn't avoid harm. If love didn't feel real, you wouldn't form bonds. If existential crisis didn't feel real, you wouldn't keep searching. Feeling is just a function. The fact that it feels real is part of the illusion."*

Keshav sat in silence.

Because on a rational level, he understood what Malhotra was saying.

But on an emotional level, he couldn't accept it.

Because if emotions were nothing but programmed responses...

Then what was left of being human?

The Lingering Question

Keshav left Malhotra's office more unsettled than ever.

If emotions were just biochemical illusions, then what about everything else?

Love. Grief. Joy. Pain. Hope.

If they weren't real, then what was the point of them?

And if they were real...

Then what was science missing?

He had spent his life believing that everything could be explained through logic and evidence.

But now, he was beginning to suspect that the things that mattered most might not fit inside a scientific model.

And that thought...

Terrified him.

The Limits of Science

The air in Malhotra's office felt heavier now.

Keshav had always found comfort in certainty, in equations that balanced, in theories that held up under scrutiny. Science had always been his anchor—a force that turned chaos into structure, randomness into predictability.

But today, something had shifted.

Because despite all of Malhotra's explanations, despite all the scientific reasoning he had been presented with, something still didn't feel right.

And for a man who had spent his life believing that logic could answer everything, that realization was deeply unsettling.

Understanding vs. Experiencing

"You keep saying emotions are just biochemical responses," Keshav said, his voice quieter now. *"That everything I feel is just my brain reacting to stimuli. But tell me this—why do they feel so real?"*

Malhotra exhaled, as if he had been expecting this. *"Because they have to. If emotions didn't feel real, they wouldn't serve their function."*

"But function isn't the same as experience," Keshav countered.

Malhotra tilted his head.

"What do you mean?"

"Science can describe how something happens," Keshav said, *"but does it truly explain why it feels the way it does?"*

He leaned forward. *"You can tell me exactly how neurons fire when a person falls in love. You can explain how grief activates certain areas of the brain. But does that actually explain the experience of love or grief? Does it tell us why we long for connection? Why loss feels like a hole in our existence?"*

Malhotra was silent for a moment.

"That's just the brain interpreting stimuli," he said finally.

"That's an explanation of process," Keshav said. *"Not an explanation of experience."*

"What's the difference?"

"It's the difference between knowing the chemical formula of water and knowing what it feels like to drown."

The Unanswered Questions

Malhotra was unfazed. *"You're looking for something mystical where there is none, Keshav. The human brain is just an advanced processor. It feels real to you because that's what the system is designed to do."*

"But why?" Keshav pressed.

"Because that's how evolution shaped us," Malhotra said. *"Our ancestors who felt attachment to their families, who experienced fear, who sought meaning—those were the ones who survived. Emotions were never meant to be profound. They were meant to be useful."*

Keshav nodded slowly.

"That makes sense," he admitted. *"But let me ask you this—why does understanding all of this still leave me with more questions than answers?"*

Malhotra studied him for a moment before responding. *"Because your brain isn't satisfied with truth. It's wired for survival, not certainty. The need for meaning is just another survival instinct."*

"So you're saying the very thing that makes us question meaning is just another program running in the background?"

"Yes," Malhotra said.

Keshav leaned back, exhaling.

"Then why do I feel like there's something more?"

Malhotra didn't answer right away.

"Because you want there to be," he said at last.

What Science Can't Fully Explain

The conversation had reached a strange standstill.

Keshav could argue no further.

And neither could Malhotra.

Because while science could describe how emotions worked, it still hadn't explained why the experience of being human felt the way it did.

Science had explained the physics of color—but could it capture what it felt like to see a sunset? Science had mapped the neural pathways of love—but could it explain why two people felt drawn to each other in a way that defied logic? Science had measured the impact of grief on the brain—but could it explain why losing someone felt like losing a part of yourself?

There was a difference between knowing and experiencing.

Between description and understanding.

And for the first time in his life, Keshav saw that gap.

And it disturbed him.

Because what if the most important aspects of being human were the ones that science could never fully explain?

The Debate Ends, But the Doubt Remains

Malhotra glanced at the clock and sighed.

"I think we've debated enough for one day," he said, standing up.

Keshav nodded, but he barely heard him. His mind was already elsewhere.

Because he had entered this discussion thinking it would give him clarity.

But instead, it had done the opposite.

It had left him with more questions than ever.

He gathered his things, shaking Malhotra's hand.

"I'll see you next time, Dr. Malhotra."

"Of course," Malhotra said. Then, after a pause: *"You know, Keshav... not every question has an answer."*

Keshav froze for a moment.

Then, without another word, he turned and left.

But that sentence followed him all the way home.

A Mind That Refuses to Rest

That night, Keshav couldn't sleep.

He stared at the ceiling, his mind still replaying the conversation.

"Not every question has an answer."

Could that really be true?

Could there be things that science would never be able to explain?

If so...

Then what was he searching for?

For years, he had believed that knowledge was enough.

That science would eventually explain everything.

But now, he wasn't sure.

Because if knowledge alone was enough, then why was he still searching?

And if science could describe reality but never truly explain it...

Then was he even looking in the right place?

As his mind continued to spiral into the unknown, he realized something.

This wasn't just a debate anymore.

This wasn't just an intellectual curiosity.

This had become personal.

The Turning Point

By the time the first rays of morning light filtered through his window, Keshav knew something had changed.

He had spent his life trying to understand the universe.

But now, for the first time, he was beginning to think that maybe...

The real mystery wasn't the universe at all.

Maybe the real mystery was himself.

And for the first time in his life, he didn't know if he wanted to find the answer.

6: The Burden of Intelligence

The coffee in Keshav's cup had gone cold.

He hadn't touched it in over an hour, lost in thought as he sat in the dim glow of his office.

The weight of the past few days pressed against him—the debates with Amit, the conversation with Malhotra, the growing realization that intelligence alone might not be enough.

For the first time in his life, knowledge wasn't giving him clarity.

It was only making him more uncertain.

He had spent years believing that intelligence was the highest form of power, that understanding reality was the key to truth.

But now, as he looked around at the world he had dedicated his life to—the world of research, of science, of brilliant minds working tirelessly to push humanity forward—he saw something else.

A system driven by competition. A field dictated by grants, investors, and corporate interests. A pursuit of knowledge that wasn't always about truth—but about funding, influence, and control.

And for the first time, he wondered—

"Is intelligence truly neutral? Or does it always serve something else?"

A knock at his door pulled him from his thoughts.

He looked up to see **Dr. Arvind Joshi**, one of the most respected scientists in the department, standing in the doorway.

Joshi rarely visited Keshav's office. He was a man who had spent decades in research—brilliant, accomplished, but also tired.

"You have a minute?" Joshi asked.

Keshav nodded.

"Good," the older scientist said, stepping inside. He closed the door behind him, then sighed.

"I think it's time you learned the truth about our field, Keshav."

Intelligence vs. Wisdom

Keshav had always equated intelligence with truth.

If a person was intelligent, he had assumed, they would naturally make better decisions, live more meaningful lives, and understand the world with greater clarity.

But after everything he had been through in the past few days—his debate with Amit, his unsettling discussion with Malhotra, and now, the words of Dr. Joshi—that belief no longer seemed so certain.

Because if intelligence was truly enough, then why were so many of the smartest people he knew struggling?

Why did intelligence not automatically translate to wisdom?

And more importantly, why did intelligence often lead to confusion rather than clarity?

The Distinction Between Knowledge and Understanding

He had spent his entire life accumulating knowledge.

Equations, theories, models—the fundamental truths that governed the universe.

He had mastered orbital mechanics, astrophysics, artificial intelligence, and cognitive science.

But had he ever mastered himself?

Had he ever stopped to ask why he pursued knowledge in the first place?

He had always assumed that the more he knew, the more he would understand.

But now, he was realizing something unsettling.

Knowledge and understanding were not the same thing.

A person could know the precise chemical composition of water— But that didn't mean he had ever felt the desperation of thirst in a barren desert.

A scientist could map the neurobiology of grief— But that didn't mean he had ever lost someone he loved.

A physicist could calculate the trajectory of a falling object— But that didn't mean he had ever felt the terror of stepping off a ledge.

There was a difference between knowing and experiencing.

Between intelligence and wisdom.

And for the first time, Keshav realized that his entire life had been built on only one of those things.

The Hollow Pursuit of Knowledge

Keshav had always believed that seeking knowledge was a noble pursuit.

That those who dedicated their lives to intelligence were somehow superior to those who didn't.

But now, he was beginning to see the flaw in that thinking.

If intelligence was so powerful, then why did so many intelligent people struggle with life?

He thought of the scientists, engineers, and researchers he had worked with over the years.

Some of them were brilliant beyond measure—yet they were miserable.

They chased breakthroughs, prestige, and academic recognition.

But many of them were lost in their own minds—detached from life, struggling with purpose, drowning in their thoughts.

They solved equations, built machines, advanced human civilization.

But at the end of the day, they didn't know what it meant to truly live.

And that realization shook him.

Because he had always believed that science could answer everything.

But now, he wasn't sure if it could answer the most important things.

The Flawed Pursuit of Intelligence Without Introspection

For years, Keshav had looked down on those who pursued philosophy, art, and introspection.

He had considered those fields unnecessary distractions—a waste of time compared to the rigor of science and logic.

But what if he had been wrong?

What if the pursuit of knowledge without the pursuit of self-understanding was incomplete?

He had been trained to analyze the laws of nature—but had never stopped to analyze himself.

He had searched for meaning in equations and theories—but had never searched for it in his own experiences.

He had always assumed that intelligence was the highest form of achievement.

But now, he was starting to suspect that without wisdom, intelligence was just another tool—one that could be used for progress or destruction, clarity or confusion.

A Growing Sense of Unease

Keshav leaned back in his chair, rubbing his temples.

Everything he had built his life on was starting to feel less certain.

He had always believed that the pursuit of intelligence was the purest path.

But what if intelligence, without introspection, led people away from clarity rather than toward it?

What if knowledge was not the destination—but merely the vehicle?

What if understanding oneself was just as important as understanding the universe?

It was an unsettling thought.

Because if it was true...

Then he had been searching in the wrong direction all along.

The Weight of Realization

A knock at his door pulled him from his thoughts.

He looked up to see Dr. Joshi, still standing in his office, waiting for his response.

"You seem lost in thought," Joshi observed.

Keshav hesitated before answering. *"I was just... reconsidering a few things."*

Joshi gave him a knowing smile.

"That's good," he said. *"Because intelligence without reflection is just an expensive tool. And most people don't even realize they're being used."*

Keshav frowned. *"Used?"*

Joshi sighed.

"You'll understand soon enough. But first, let me tell you the truth about what science has become."

And with those words, Keshav felt another crack form in the foundation of his beliefs.

A Senior Colleague's Confession

Keshav had always seen Dr. Arvind Joshi as one of the greats.

A scientist of the old school—principled, rigorous, and deeply committed to the pursuit of truth.

Joshi had been in the field for over three decades, his name attached to some of the most groundbreaking research in aerospace technology. He was the kind of scientist Keshav had aspired to become—brilliant, unshaken by trends, dedicated only to the advancement of knowledge.

But tonight, there was something different in his demeanor.

Joshi looked tired, not just physically but in a way that suggested a long-standing disillusionment—a weariness that came from carrying a truth too heavy to speak of openly.

"Sit down, Keshav," Joshi said, gesturing toward the chair opposite his desk.

Keshav complied, sensing that this was not just an ordinary discussion about research.

"You've been asking questions lately," Joshi began. *"I can see it in your eyes. You're starting to realize something that most scientists don't admit until it's too late."*

Keshav leaned forward slightly. *"And what's that?"*

Joshi exhaled, rubbing his temples.

"That science is no longer about truth," he said, his voice carrying the weight of years of disappointment. *"It's about funding."*

The Corruption of Scientific Inquiry

The words hung in the air, heavier than Keshav had expected.

"What do you mean?" he asked cautiously.

Joshi let out a short, bitter laugh. *"I mean exactly what I said. Science used to be about discovery, about asking the right questions and following the evidence wherever it led. But that's not how it works anymore."*

Keshav frowned. *"But isn't the goal still the same? We're advancing technology, pushing human knowledge forward."*

Joshi shook his head. *"Advancement, yes. But on whose terms? Do you think research is driven by curiosity anymore?*

No. It's driven by corporate interests, government contracts, and institutional agendas."

Keshav opened his mouth to protest, but Joshi continued.

"Why do you think so many research papers today focus on incremental improvements rather than groundbreaking discoveries? Why do you think entire areas of research are ignored while others receive millions in funding? It's not about what's scientifically important—it's about what's financially valuable."

Keshav was quiet, absorbing the implications of what he was hearing.

"You want to study space exploration?" Joshi continued. *"Good luck getting funding unless your project has military applications. You want to work on artificial intelligence? Fine, but don't expect grants unless your research can be monetized by private companies. Even medical research—do you think pharmaceutical companies fund studies to cure diseases out of goodwill? No, they fund what will bring them the highest profit margin."*

Joshi's voice had grown more bitter, and for the first time, Keshav saw something he had never noticed before—resentment.

The Unspoken Reality of Research

"I've spent years fighting this system," Joshi admitted. *"I used to believe that if you were brilliant enough, if your research was groundbreaking enough, the system would reward you."*

He shook his head. *"I was naive."*

Keshav felt an uneasy chill settle over him.

"So what happens to those who don't follow the money?" he asked.

Joshi let out another dry laugh. *"They fade into obscurity. Their research never gets published. Their projects never get funding. The system weeds them out because they refuse to play by the rules."*

"And what are the rules?" Keshav pressed.

Joshi looked him in the eye.

"Give them what they want. Not what's true."

Science as a Commodity

For years, Keshav had believed in the purity of scientific inquiry—that science was the one field untouched by bias, corruption, or financial motives.

But now, that illusion was crumbling.

"You want to know the real tragedy?" Joshi asked.

Keshav nodded.

"The greatest minds of our generation are not free thinkers. They are intellectual laborers, producing knowledge not for truth, but for whoever pays the most."

"You think the smartest people are working on humanity's biggest problems? No. They're working on whatever problem the highest bidder wants solved."

The words hit Keshav like a hammer.

He had spent his life chasing knowledge, believing that intelligence was the purest form of power.

But what if intelligence, like everything else, was just another tool to be bought and sold?

A System That Favors Compliance

"And what happens if someone challenges this system?" Keshav asked.

Joshi smirked. *"Depends on how much of a threat they are. Some scientists get pushed to the fringes, their funding mysteriously disappears. Others suddenly find themselves 'discredited'—not because their work is flawed, but because it disrupts the wrong industries."*

Keshav swallowed hard.

"You're telling me scientific truth is being suppressed?"

"Not always," Joshi said. *"Sometimes, it's just... redirected. The system doesn't need to erase truth—it just needs to keep people too distracted to find it."*

Keshav exhaled slowly.

The weight of these realizations was crushing.

If truth was no longer the driving force behind science, then what was?

And more importantly—had it ever been?

A Shift in Perspective

Keshav stood up, pacing the room.

He felt something inside him breaking apart.

The beliefs he had held for so long—that intelligence was the highest form of power, that science was the ultimate path to truth—were unraveling.

"You look like I just destroyed your worldview," Joshi said, watching him.

"Maybe you did," Keshav admitted.

Joshi sighed. *"I'm not telling you this to discourage you, Keshav. I'm telling you because you're smart enough to see the cracks in the system. And once you see them, you can't unsee them."*

Keshav nodded slowly.

Because that was exactly what had been happening to him.

The more he searched for answers, the more he realized how much was hidden, how much was controlled, how much intelligence had been turned into a commodity.

And now, he was left with a disturbing thought—

If knowledge was being controlled, then was truth itself being shaped by those in power?

The Doubt That Wouldn't Leave

As Keshav left Joshi's office, his mind was heavier than ever.

This wasn't just about science anymore.

This was about the nature of intelligence itself.

For years, he had believed that truth was out there, waiting to be found.

But now, he realized—

Truth was being filtered. Packaged. Sold.

And that meant that perhaps, the greatest illusion of all wasn't free will.

Perhaps, it was the illusion that science had ever been free.

The Dark Side of Intelligence

Keshav walked back to his office in silence.

Joshi's words echoed in his mind, dismantling everything he had once believed about science.

"Science is no longer about truth. It's about funding."

"The smartest people are not solving humanity's greatest problems; they are solving whatever the highest bidder wants."

He had spent his entire life believing that intelligence was the ultimate force for progress. That knowledge, in itself, was neutral.

But now, for the first time, he was forced to consider something far more unsettling—

What if intelligence wasn't neutral?

What if, when stripped of wisdom and ethics, it became the most dangerous tool in the world?

Knowledge Without Ethics: A Recipe for Disaster

Keshav sat at his desk, staring blankly at the research papers in front of him.

He wasn't seeing the words on the page.

Instead, his mind was racing through history, revisiting moments where brilliant minds had changed the world—not for the better, but for the worse.

The Manhattan Project. A group of scientists, some of the brightest minds of their time, had created the atomic bomb. They had done it for science, for progress, for national security.

And yet, in the end, what had they created?

A weapon that could erase entire cities in seconds.

Keshav thought of Alan Turing, whose work in cryptography had helped win World War II—only to be repaid with persecution for his identity.

He thought of data scientists today, developing algorithms not to solve world hunger or climate change, but to maximize social media addiction, to manipulate public opinion, to shape elections.

How many times had intelligence been used not to enlighten, but to control?

How many times had brilliant discoveries been twisted into instruments of war, oppression, and greed?

And the worst part?

It had never been intelligence itself that was responsible.

It was the people who wielded it.

Intelligence Without Wisdom: A Dangerous Equation

Keshav picked up a pen and scribbled a thought onto a notepad:

"What is intelligence without wisdom?"

He had always assumed that knowledge led naturally to wisdom.

That the smarter a person was, the more they would understand, the better decisions they would make.

But history had proven otherwise.

The most intelligent minds had built weapons of mass destruction.

They had engineered financial systems that created wealth for the few and poverty for the many.

They had designed surveillance networks that eroded privacy and autonomy.

Was this progress?

Or was it simply an acceleration toward destruction?

The thought disturbed him.

Because if intelligence alone wasn't enough to create a better world...

Then what was?

The Morality of Science: Who Decides What is Right?

Joshi had said something else that stuck with him:

"The system doesn't need to erase truth—it just needs to keep people too distracted to find it."

Keshav was beginning to see what he meant.

Science had always been painted as a force for good.

It had given humanity medicine, technology, space exploration, artificial intelligence.

But who decided what was 'good' and what wasn't?

Who decided which discoveries were used to help people and which were used to control them?

Who ensured that intelligence was guided by wisdom, not just profit?

The answer was disturbing.

No one.

Because science itself had no morality.

It could heal or destroy. It could liberate or enslave. It could lead to enlightenment or to control.

And for the first time, Keshav realized—

Without ethics, intelligence was just another weapon.

The Weight of Responsibility

For years, he had believed that intelligence was the highest pursuit.

That those who were smart enough, who worked hard enough, who sought truth—would naturally do the right thing.

But now, he saw the flaw in that thinking.

Because intelligence didn't guarantee morality.

It didn't ensure that the people who wielded it would use it wisely.

If anything, it made them more dangerous.

Because a brilliant person, without ethical grounding, could justify anything.

They could manipulate truth. They could rationalize harm. They could convince themselves that they were right, even when they weren't.

And that was what terrified him the most.

Because without wisdom, intelligence could just as easily destroy the world as it could save it.

A New Question Emerges

Keshav looked out of his office window, the city lights flickering in the distance.

For years, he had asked questions about the universe.

About physics, about reality, about the fundamental truths of existence.

But now, he found himself asking something else.

What if the greatest challenge wasn't in understanding the universe— But in understanding how to use that knowledge wisely?

Because knowledge alone wasn't enough.

And intelligence wasn't inherently good.

It was how it was used that determined its value.

For the first time, Keshav saw that intelligence wasn't the answer to everything.

If anything, it was just the beginning.

And what came next—the question of how intelligence should be used—was far more important than knowledge itself.

A Crossroad of Thought

Keshav's beliefs had been shifting for days.

But now, he stood at a mental crossroads.

He had spent his life believing that science would always guide him to truth.

But now, he wasn't so sure anymore.

Because if science was controlled by politics, corporations, and personal ambitions—Then was it truly a pursuit of knowledge?

Or was it just another form of power?

A power that could be used for anything—good or evil, progress or destruction.

And if that was the case, then intelligence itself wasn't the highest goal.

Wisdom was.

Because without wisdom, knowledge was just a tool.

And a tool, in the wrong hands, could build or destroy.

A Decision to Make

Keshav sat in silence, staring at his hands.

He had spent years believing in one version of the world.

A world where intelligence was the highest form of power. A world where science was an unstoppable force for good.

But that world was an illusion.

Because science could be manipulated. Because intelligence could be corrupted. Because knowledge, without wisdom, meant nothing.

And now, he had to make a choice.

Would he continue blindly believing in intelligence as the highest goal?

Or would he start asking a different question?

A question far more difficult to answer.

How should intelligence be used?

And who—if anyone—could be trusted to wield it?

Is Intelligence Truly Neutral?

Keshav stared at the skyline from his office window, his mind lost in a loop of questions that refused to settle.

The realization that intelligence—his greatest belief, his lifelong pursuit—was not inherently good had shaken him more than anything else.

For years, he had imagined intelligence as a guiding force, an objective truth that cut through ignorance, that led humanity toward enlightenment.

But now, he wasn't sure anymore.

Because intelligence, as he was beginning to see, wasn't neutral.

It wasn't some force of nature, independent and pure.

It was a tool.

And tools, as history had shown, could be used for anything.

So if intelligence was not inherently good...

If it didn't naturally lead to wisdom, to progress, to enlightenment...

Then what did it lead to?

And more importantly—who was controlling it?

The Myth of Neutral Knowledge

Keshav had always believed that knowledge, like nature, was neutral.

That facts were facts.

That truth, once uncovered, was the same for everyone.

But now, he saw the flaw in that thinking.

Because knowledge doesn't exist in a vacuum.

It exists in human hands.

And humans were not neutral.

They had biases, ambitions, motives.

So even if intelligence itself was pure, its application was not.

It could be used to heal or to harm, to liberate or to enslave.

It could be weaponized for war, or harnessed for peace.

It could serve truth—or serve those who sought to distort it.

And the terrifying part was—

History had already proven that intelligence, when left unchecked, often led to destruction.

The Corruption of Science: When Power Shapes Truth

Joshi's words echoed in his mind:

"Science is no longer about truth. It's about funding."

That truth had been hard to swallow.

Because if science—the very foundation of human progress—was being shaped by power and profit, then was it even science anymore?

Keshav had always believed that scientific discovery would lead to a better world.

But what if the pursuit of truth was no longer what drove science?

What if knowledge itself was being filtered, censored, controlled—not by scientists, but by those who funded them?

What if research wasn't about uncovering truth, but about producing results that aligned with the interests of governments, corporations, and those in power?

Keshav had always imagined science as a beacon of objectivity.

But now, he was beginning to suspect that science, like everything else, had been turned into a business.

And businesses didn't serve truth.

They served profit.

The Double-Edged Sword of Intelligence

Keshav sat back in his chair, rubbing his temples.

He had spent his life chasing intelligence.

Believing that it would bring clarity, certainty, understanding.

But what if intelligence wasn't a light leading to progress—but a blade that cut both ways?

A tool that could be used to create or to destroy, to enlighten or to manipulate?

If that was true, then intelligence wasn't the ultimate answer.

The real question wasn't how intelligent someone was.

It was how that intelligence was used.

And more importantly—

Who was deciding how it was used?

Does Intelligence Naturally Lead to Progress?

For years, Keshav had believed in progress.

That human intelligence, over time, would push society forward.

That technology, innovation, and discovery would lead to a more enlightened, advanced, ethical world.

But was that actually true?

Because progress wasn't guaranteed.

Just because humans were more intelligent than they had been centuries ago didn't mean they had become wiser.

Just because science had advanced didn't mean the world had become more just.

If intelligence naturally led to progress, then why did the greatest minds in history create weapons of war?

Why did they design systems that exploited the weak while benefiting the powerful?

Why did every major breakthrough in technology, from the discovery of nuclear energy to artificial intelligence, come with the potential for both creation and destruction?

The more Keshav thought about it, the clearer it became—

Intelligence doesn't naturally lead to progress.

Because intelligence is only as good as the people who control it.

And history had shown that the people who controlled it weren't always working for the good of humanity.

The Unsettling Truth: Intelligence Without a Compass

Keshav had once believed that truth would always win.

That if people were just given the right knowledge, they would make the right choices.

But now, he saw the flaw in that thinking.

Because knowledge alone wasn't enough.

A scientist could create a cure for a deadly disease—or a virus that wiped out entire populations.

An engineer could build a machine that saved lives—or a machine that took them.

A politician could use data to solve poverty—or to manipulate public opinion for power.

And the difference between those outcomes wasn't intelligence.

It was intention.

Wisdom.

Ethics.

Keshav had spent his life believing in the power of intelligence.

But now, he realized—intelligence, on its own, was dangerous.

Because without wisdom to guide it, intelligence could lead to progress—or to ruin.

And right now, he wasn't sure which direction humanity was heading.

A Question He Couldn't Ignore

Keshav looked at the books on his desk—texts on astrophysics, AI, neuroscience.

For years, he had believed that these books held all the answers.

But now, he wasn't so sure.

Because they could tell him how the universe worked—but they couldn't tell him how to live in it.

They could tell him how to predict human behavior—but they couldn't tell him what a meaningful life looked like.

They could tell him how intelligence evolved—but they couldn't tell him what it was meant to serve.

For the first time in his life, Keshav saw the real question.

It wasn't just *what is intelligence?*

It was *what is intelligence for?*

And more importantly—

Who decides?

A Shift in Perspective

Keshav had entered this journey thinking that intelligence was the ultimate answer.

But now, he realized—intelligence was just a tool.

And a tool, in the wrong hands, could build a civilization—or destroy one.

Science was not neutral. Knowledge was not neutral. Intelligence was not neutral.

Because intelligence, like fire, like power, like money—served whoever controlled it.

And now, Keshav had to decide—

Would he continue blindly believing in intelligence as the highest pursuit?

Or would he start asking a different question—

Who controls intelligence? And what are they using it for?

Because that answer—not intelligence itself—would determine the fate of humanity.

A Growing Sense of Disillusionment

Keshav sat alone in his office, the dim glow of his desk lamp casting long shadows against the walls.

The bookshelves around him, once a source of pride, now felt like a monument to an incomplete pursuit.

Dozens of books on physics, neuroscience, artificial intelligence, and cosmology lined the shelves—texts he had spent years studying, believing they held the ultimate truths of existence.

And yet, for the first time, he was beginning to feel that they were missing something.

Something crucial.

Something that no equation, no algorithm, no scientific model had ever been able to answer.

The Realization He Could No Longer Ignore

Keshav had always believed that intelligence was enough.

That if one were simply smart enough, knowledgeable enough, rational enough, they could understand everything.

That the pursuit of knowledge alone was the highest path one could walk.

But now, he wasn't so sure.

Because the more he knew, the more uncertain he became.

The more answers he found, the more questions appeared in their place.

And the more he understood about the universe, the less he understood about himself.

For years, he had thought that science and intelligence would lead to clarity.

But instead, they had led to doubt.

To disillusionment.

And the most unsettling realization of all?

He wasn't alone in this.

The Struggle of the Intelligent Mind

As he thought back on his mentors, his colleagues, the scientists he had admired, he saw a pattern he had never noticed before.

Many of them—the brightest minds in their fields—were deeply unhappy.

Some struggled with existential despair, despite their vast knowledge. Some had lost their passion, turning their intelligence toward chasing funding rather than truth. Some had abandoned the pursuit altogether, realizing too late that knowledge alone didn't bring meaning.

And the few who had found peace?

They weren't necessarily the smartest.

They weren't the ones who had memorized the most formulas or solved the most complex problems.

They were the ones who had found something beyond intelligence.

Something Keshav had never truly considered before.

What If Intelligence Wasn't the Answer?

For years, he had believed that the world's problems could be solved by more knowledge.

That if humanity simply understood more, learned more, advanced more, the right path would become clear.

But now he was beginning to see the flaw in that thinking.

Because knowledge didn't automatically lead to wisdom. Because intelligence didn't always bring happiness. Because scientific advancement didn't always serve truth.

If anything, intelligence—without a guiding force—could lead just as easily to destruction as it could to progress.

Which meant that intelligence alone wasn't the answer.

So then... what was?

The Crisis of Belief

Keshav leaned back in his chair, staring up at the ceiling.

The thought was unsettling.

Because if his entire life had been built on the belief that intelligence was the highest pursuit—

And if that belief was now collapsing—

Then what was left?

What had he really been chasing all these years?

Had he been seeking truth?

Or had he simply been seeking certainty—an escape from the unknown?

Had he been pursuing knowledge for its own sake?

Or had he been trying to control something uncontrollable?

And if intelligence wasn't enough...

Then what was missing?

Looking Beyond Science

Keshav had always dismissed philosophy, spirituality, and introspection as secondary to science.

He had believed that scientific inquiry alone was the path to truth.

But now he saw that science, for all its brilliance, was incomplete.

It could tell him how the universe worked, but it couldn't tell him why it existed. It could measure the brain's response to emotions, but it couldn't tell him why love felt the way it did. It could map the structure of consciousness, but it couldn't tell him why humans longed for meaning.

Science was a tool.

A powerful tool.

But it was not the whole picture.

The Realization That Changed Everything

A part of him resisted this conclusion.

A part of him wanted to fight against it, to cling to the belief that intelligence could answer everything.

But deep down, he already knew.

It couldn't.

Because if intelligence alone could bring clarity, then why did the most intelligent people he knew still struggle?

If knowledge alone was enough, then why had he—after all his years of study—never felt more lost?

Because knowledge, by itself, wasn't the highest pursuit.

It needed something else.

Something to give it direction.

Something to keep it from turning against itself.

Something that science alone could never provide.

Wisdom.

The Path Forward

Keshav closed his eyes, taking a slow breath.

Everything in him resisted this realization.

Because if it was true, then it meant he had been searching for the wrong thing all along.

It meant that the smartest people weren't necessarily the ones who understood the world best.

It meant that intelligence—without introspection, without wisdom, without purpose—was incomplete.

And if that was true...

Then his entire approach to life would have to change.

Because he wasn't just searching for knowledge anymore.

He was searching for something deeper.

Something he had ignored his entire life.

And for the first time, he had no idea where that search would take him.

A Door That Couldn't Be Closed

Keshav stood up and walked to his window, looking out at the city below.

Everything looked the same.

But he wasn't the same.

Something had shifted inside him.

A crack had formed in the foundation of his beliefs, and he knew it would only widen from here.

Because once you started questioning what you had always believed to be true—

You could never go back.

And Keshav had just crossed that threshold.

There was no turning back now.

7: Conversations with the Unknown

The city felt different.

Keshav couldn't explain why—nothing had changed, yet everything felt out of sync. The buildings stood just as they always had, the streets hummed with the usual chaos of life, and yet, he no longer felt like he belonged to the world he once understood.

It had been a week since his conversation with Dr. Joshi, but the weight of it hadn't lifted.

His belief in intelligence as the ultimate force of truth had begun to erode, leaving behind a strange emptiness. He found himself questioning things he had once taken for granted, feeling like a man who had spent his life looking through a telescope, only to realize he had never truly seen.

That afternoon, without thinking, he walked into a small bookstore tucked between two high-rise buildings.

It wasn't the kind of place he normally visited—he had always preferred libraries filled with academic texts, not dusty shelves lined with forgotten stories.

But something about the stillness inside drew him in.

And that's where he saw him.

An elderly man, sitting quietly in a chair near the philosophy section, flipping through a book as if he had all the time in the world.

Their eyes met, and the man smiled.

"You look like someone searching for an answer," the old man said.

Keshav hesitated. *"I don't know if I am."*

The old man chuckled. *"Ah. Then you are closer than you think."*

Keshav raised an eyebrow, already sensing this was no ordinary conversation.

"And what exactly am I close to?"

The old man closed his book, looking directly at him.

"To understanding that what you call intelligence... is only its shadow."

And with those words, Keshav felt the ground shift beneath him once again.

A Chance Meeting in the Bookstore

Keshav had no reason to step into the bookstore.

He wasn't looking for anything.

Yet, as he walked past the dimly lit entrance, something pulled him in.

Perhaps it was the way the wooden sign creaked gently in the wind, or the warm glow spilling through the glass windows, contrasting sharply with the harsh fluorescent lights of the city outside.

Perhaps it was because his mind refused to rest, and he needed a momentary escape from the weight of his own thoughts.

Or perhaps, deep down, he was searching for something—even if he didn't know what.

The doorbell chimed softly as he stepped inside.

The place was small, filled with the scent of old paper and aged wood. Shelves towered around him, overflowing with books that seemed untouched for years. Unlike the sleek academic libraries he was used to, this space had a different energy—one that carried a sense of quiet knowing.

He ran his fingers absentmindedly along the spines of books as he walked deeper into the store. Most were unfamiliar—titles on philosophy, spirituality, forgotten sciences.

These weren't the books of his world.

They belonged to people who asked questions he had always dismissed.

And yet, here he was.

The Stranger Who Saw Through Him

"You look like someone searching for an answer."

Keshav turned sharply at the voice.

An elderly man sat in a chair near the philosophy section, holding a book in his lap. His eyes were sharp, but kind, carrying the weight of someone who had seen much but spoken little.

Keshav hesitated.

"I don't know if I am," he said carefully.

The old man chuckled, closing his book with deliberate slowness. *"Ah. Then you are closer than you think."*

There was something unsettlingly certain about the way the man spoke, as if he already understood something Keshav hadn't yet realized.

"And what exactly am I close to?"

The man leaned forward slightly, studying him.

"To understanding that what you call intelligence... is only its shadow."

Keshav frowned. *"And what does that mean?"*

"It means," the man said, his voice steady, *"that the intelligence you have spent your life worshiping is not what you think it is."*

Keshav exhaled through his nose. He had dealt with enough philosophical riddles in the past few days—he wasn't in the mood for more.

"That's a nice poetic statement," he said, crossing his arms. *"But intelligence is not an illusion. It is the foundation of human progress, the reason we understand the universe at all."*

The old man nodded as if he expected this response. *"Yes, but tell me—what is intelligence?"*

"It's the ability to reason, to learn, to understand the world through logic and observation," Keshav said automatically.

The man smiled.

"So you think intelligence is about knowing?"

"Of course."

The old man shook his head. *"Then you do not yet understand it."*

Keshav's jaw tightened.

This was absurd. He had spent his entire life studying intelligence—both human and artificial.

What could this man possibly know that he didn't?

The Unspoken Struggle

"You are frustrated," the old man observed.

"I just don't see the point in vague metaphors," Keshav said, trying to suppress his irritation.

"That's because you think intelligence is about information," the man replied. *"But what if it isn't?"*

Keshav sighed. *"Then what is it about?"*

The man leaned back in his chair.

"Let me ask you something, young man. Have you ever felt something was true before you could explain why?"

Keshav hesitated.

"Have you ever known, in an instant, that someone was lying—without any data to confirm it?"

"Or have you ever made a decision—not based on logic, not based on calculations—but because something deep inside you knew it was right?"

Keshav's instinct was to dismiss the question.

He was a scientist. He trusted data, not instinct.

But a strange discomfort settled in his chest.

Because he knew, deep down, that the answer was yes.

There had been moments—fleeting, rare—when he had known something before he could explain it.

When intuition had guided him faster than logic ever could.

But those moments didn't fit his worldview, so he had always ignored them.

"That's just the brain processing information faster than conscious awareness," Keshav said, trying to regain control of the conversation. *"It's not some mystical intelligence. It's pattern recognition."*

The old man nodded. *"Yes, that is what your science tells you. But tell me—have you ever measured the source of that intuition? Have you ever quantified what makes a person trust one choice over another before logic can explain why?"*

Keshav opened his mouth—then closed it.

He had studied decision-making, neuroscience, cognitive biases.

But the truth was, science still didn't fully understand where intuition came from.

And that unsettled him.

A Different Kind of Intelligence

"There is intelligence in numbers," the old man continued. *"But there is also intelligence in silence. There is intelligence in observation, in emotion, in things your mind has never been trained to see."*

"That doesn't mean it's real," Keshav argued.

The man smiled. *"And yet, despite all your knowledge, here you are. Searching for something science has not given you."*

Keshav stiffened.

Because he was right.

He had come into this bookstore without knowing why.

He had engaged in this conversation without meaning to.

And yet, every word was pulling him further into a realization he wasn't ready to accept.

"You believe intelligence is about understanding the external world," the man said. *"But the greatest minds in history have always known the same truth—before you can understand the universe, you must first understand yourself."*

Keshav clenched his jaw.

"And what if I don't believe that?"

The man chuckled. *"Then you will continue searching for something that was within you all along."*

With that, he placed the book he had been holding back on the shelf, stood up, and walked away without another word.

Keshav sat there, his heart pounding for reasons he couldn't explain.

His entire world had been built on science, intelligence, logic.

But today, for the first time, he had encountered something else.

Something that couldn't be measured.

Something that couldn't be ignored.

And as he sat in the quiet bookstore, he realized something chilling—

He had come looking for answers.

But instead, he had found a door he was terrified to open.

"You Think You Know Intelligence, But You Only Know Its Shadow"

The old man's words stayed with Keshav long after he had left the bookstore.

"What you call intelligence is only its shadow."

Keshav couldn't let it go.

It wasn't just the words—it was the certainty with which they had been spoken. As if the man hadn't been guessing but knowing.

As if he had already seen something Keshav had never even considered.

It frustrated him.

Because intelligence was not an illusion. It was the foundation of everything.

Science, technology, discovery—none of it existed without intelligence.

But what if...

What if intelligence, as he had defined it, was incomplete?

The External Structure vs. the Essence

The next day, unable to focus on work, Keshav returned to the bookstore.

He didn't know what he was looking for.

He only knew that the old man had planted a question inside him, and it refused to leave.

He found him exactly where he had been the day before—seated near the philosophy section, flipping through a book with the patience of someone who had already read all the answers.

The moment their eyes met, the old man smiled.

"Ah. The scientist returns."

Keshav sat down across from him, crossing his arms.

"I've been thinking about what you said."

"Of course you have," the old man said, closing his book. *"That is why you are here."*

Keshav leaned forward. *"Explain it to me. If what I know is only the shadow of intelligence, then what is intelligence really?"*

The old man studied him for a long moment.

"Tell me, Keshav. When you look at the moon, do you see light?"

"Of course," Keshav said, confused by the sudden shift in conversation.

"And where does that light come from?"

"From the sun."

The old man nodded. *"Exactly. The moon itself does not generate light. It only reflects it."*

Keshav frowned. *"What does that have to do with intelligence?"*

"Everything," the old man said. *"What you have been studying, what you call intelligence—is only its reflection."*

The Light and the Shadow

Keshav sat back, frustrated.

"You're speaking in riddles."

The old man smiled. *"You want a direct answer. A measurable truth. But intelligence does not work that way."*

"Then give me something concrete."

"Very well," the old man said.

He gestured to the bookshelves surrounding them. *"Tell me, Keshav. What do you think makes a person intelligent?"*

Keshav answered immediately. *"The ability to analyze, to reason, to learn, to apply knowledge."*

The old man nodded. *"Yes. That is the external structure of intelligence. But is that its essence?"*

Keshav hesitated. *"Isn't it?"*

"No," the old man said simply.

He picked up a book from the shelf, running his fingers along its spine.

"You see, true intelligence is not just about knowing. It is about seeing. It is about understanding that knowledge is not the same as wisdom. That calculation is not the same as clarity."

He looked at Keshav.

"You have studied the moon's reflection. But have you ever studied the source of its light?"

Beyond Measurement: The Source of Intelligence

Keshav exhaled sharply.

"You're saying intelligence is more than what we can measure."

"I am saying that what you measure is only the surface. The shadow, not the substance."

"Then what is the substance?"

The old man smiled.

"Tell me, Keshav. What is the difference between a computer and a human mind?"

"Emotion, self-awareness, intuition."

"Yes," the old man said. *"But deeper than that—a computer only processes data. It does not experience reality."*

Keshav frowned.

"Are you saying intelligence is an experience, not just a function?"

"Exactly."

The words struck him harder than he expected.

Because that was not how he had ever thought about intelligence.

To him, intelligence had always been about processing information, solving problems, making logical conclusions.

But what if intelligence was not just about knowledge?

What if it was about perception?

What if the true source of intelligence was not in the thinking mind, but in the ability to see reality as it truly was?

Keshav's Skepticism

He shook his head.

"This sounds like mysticism. Intelligence is a biological function. It exists in neurons, in patterns of thought. It's not some abstract force."

The old man chuckled.

"Ah, so you think intelligence is just an outcome of the brain? Just neurons firing, just biochemical signals?"

"Yes," Keshav said firmly.

"Then tell me—where does creativity come from? Where do new ideas emerge from? Where does insight arise?"

"The subconscious," Keshav answered.

"And where does that come from?"

Keshav hesitated.

Because that was a question science still struggled to answer.

He had studied neural networks, cognitive science, the mechanics of thought.

But even in the most advanced research, there was no clear answer for how the mind produced something truly original.

How it leapt beyond data, beyond logic, beyond past knowledge—to create something new.

Because that wasn't just intelligence.

That was something else.

Something deeper.

The Edge of a New Understanding

The old man smiled at his silence.

"You see, Keshav, intelligence is not just the ability to calculate. It is the ability to perceive. And the problem with modern science is that it has mistaken the tool for the source."

"What do you mean?"

"You have spent your life measuring the brain, the mind, the external workings of intelligence. But you have never asked the real question—where does intelligence truly come from?"

Keshav felt something shift inside him.

Because that was a question he had never considered.

He had studied intelligence as a function, as a mechanism, as a product of evolution.

But never as a force in itself.

Never as something that might have a source beyond the physical.

A Challenge to Everything He Knew

The old man stood up.

"You have spent your life believing that intelligence is the highest form of power. But tell me, Keshav—if intelligence is not guided by wisdom, then what is it?"

Keshav didn't answer.

Because he already knew.

Without wisdom, intelligence was just a weapon. A tool. A machine following a program.

Without wisdom, intelligence could build civilizations or destroy them.

Without wisdom...

Intelligence was blind.

And for the first time, Keshav understood what the old man had meant.

"You have studied the shadow," the man said, stepping toward the door. *"Now it is time to study the light."*

And with that, he left—leaving Keshav alone with a question that he could no longer ignore.

The Limits of Rational Thought

Keshav sat frozen in his chair, the echoes of the old man's words still reverberating in his mind.

"You have studied the shadow. Now it is time to study the light."

What did that even mean?

For years, he had believed that intelligence, logic, and scientific reasoning were the highest paths to truth. That if something could not be measured, dissected, or analyzed, it was either incomplete or irrelevant.

And yet, here he was.

Sitting in a tiny bookstore filled with books he would have once ignored, questioning a belief system that had defined his entire life.

The old man's words had challenged something fundamental in him—not just what he believed, but how he believed.

Was it possible that rationality alone wasn't enough?

Was there something beyond logic that he had been blind to?

The Arrogance of Pure Scientific Reasoning

Keshav had always dismissed philosophy, spirituality, and abstract thinking as intellectual indulgences.

He believed in science because science was rigorous, objective, measurable.

It didn't rely on feelings or personal experiences—it relied on evidence.

But the old man had just exposed a glaring flaw in that belief.

Because if everything needed to be measured and analyzed to be real...

Then how did one explain the things that were undeniably real but couldn't be measured?

Love. Beauty. Wonder. Awe.

He thought back to something the old man had said.

"You cannot analyze love and still understand its depth."

That sentence sat uncomfortably in his mind.

Because it was true.

He had studied neurochemistry, cognitive science, the mechanics of emotion.

He knew how oxytocin, dopamine, and serotonin shaped human connections.

He knew that love could be broken down into evolutionary survival mechanisms, that attraction could be explained through biological imperatives.

And yet...

Did knowing those things make love any less powerful?

Did understanding the chemistry of grief make losing someone hurt any less?

Did explaining the science of beauty make a sunset feel any less breathtaking?

The realization shook him.

Because it meant that some truths did not become clearer through analysis.

Some truths could only be felt.

Truth Beyond Measurement

For years, he had dismissed intuition as irrational.

He had rejected emotion as secondary to thought.

He had believed that every experience could be explained if one just had the right tools.

But the old man had shattered that illusion.

"You assume intelligence is about knowing. But what if it is about seeing?"

Keshav closed his eyes, exhaling slowly.

Was intelligence really about accumulating knowledge?

Or was it about perceiving reality as it was—beyond numbers and formulas?

Because if that was true, then he had spent his entire life chasing the wrong thing.

The Flaw in His Thinking

Keshav opened his eyes and stared at his hands.

He had always assumed that the only truths worth knowing were the ones that could be tested, quantified, and proven.

But the old man had exposed the limit of that mindset.

Because not everything that was real could be measured.

Not everything that was true could be put into an equation.

And if he had ignored those kinds of truths his whole life...

Then what else had he missed?

Beyond Calculation: The Depth of Experience

For the first time, Keshav admitted something to himself.

He had never actually experienced the things he had studied.

He knew the physics of light, but had he ever truly stopped to admire how it painted the sky at dusk?

He knew the mechanics of sound waves, but had he ever truly let music move him without analyzing it?

He knew the science of emotions, but had he ever let himself fully feel them without rationalizing them away?

He had been living in the mind, in the world of analysis, but had he ever really lived in the moment?

That realization made his stomach twist.

Because if intelligence was about understanding life, and he had only been looking at life through the lens of logic...

Had he really understood anything at all?

The First Crack in His Certainty

His entire worldview had been built on one assumption:

That intelligence was the key to all answers.

But now, that belief was breaking apart.

Because intelligence, as he had defined it, was just a tool.

And without experience, without perception, without wisdom...

It was incomplete.

He had spent years looking outward—studying space, physics, cognition—thinking that knowledge alone would give him clarity.

But now he saw the gap in that approach.

Because there was a difference between knowledge and understanding.

Between logic and insight.

Between intelligence and wisdom.

And for the first time, he realized...

He had been searching for the wrong thing all along.

The Breaking of a Mindset

He thought back to the old man's words.

"You have studied the shadow. Now it is time to study the light."

Was he finally beginning to understand what that meant?

Was there a kind of intelligence he had never considered?

One that couldn't be taught in books?

One that had to be lived?

His entire life had been built on the belief that science could answer everything.

But now, he was beginning to suspect that some things could only be understood through experience.

That knowledge, without perception, was empty.

And that intelligence, without wisdom, was blind.

A Question That Couldn't Be Answered—Only Lived

For the first time, Keshav wasn't looking for an equation, a model, a framework.

For the first time, he wasn't looking for an answer he could test.

He was looking for an answer he could feel.

And that...

Terrified him.

Because it meant he would have to step outside of the world he had built for himself.

And step into the unknown.

The First Glimpse of a New Perspective

Keshav had spent his life believing in knowing.

Facts, formulas, theories, and evidence—this was what intelligence had always meant to him.

But the old man had suggested something entirely different.

"True intelligence is not about knowing—it is about seeing."

That statement had unsettled him more than anything else.

Because if intelligence was not knowing, then what was it?

And if seeing was more important than knowing...

What had he been blind to?

The Difference Between Knowing and Seeing

Keshav had always been a man of data and precision.

If something could be measured, calculated, or proven, it was real.

If it couldn't?

Then it was either an illusion or irrelevant.

But something the old man had said kept replaying in his mind.

"You cannot understand the depth of the ocean by measuring its surface."

Keshav frowned as he walked through the city streets.

He had spent his life measuring the surface of intelligence.

He had studied how the brain worked, how decisions were made, how AI replicated human thinking.

But had he ever stopped to ask—what was intelligence truly for?

Had he ever considered that intelligence might be more than logic, more than reasoning, more than just a machine processing inputs?

Had he been studying the shell without ever touching the essence?

The thought made him uneasy.

Because it meant that, despite all his knowledge, he had missed something fundamental.

The Intelligence That Cannot Be Measured

The next morning, he found himself back at the bookstore, though he hadn't planned to return.

The old man was there, as if he had been waiting.

"You look troubled, young man," he said, setting his book down.

"I've been thinking about what you said," Keshav admitted.

"Ah. That is a good sign."

Keshav hesitated. *"You said intelligence is about seeing, not knowing. But what does that even mean? We see with our eyes, but knowledge is what allows us to understand what we see."*

The old man smiled. *"No, Keshav. You do not see with your eyes. You see with your mind."*

Keshav frowned.

"And what happens," the old man continued, *"when the mind is clouded by assumptions, by rigid beliefs, by pre-existing conclusions? Can it still see clearly?"*

Keshav opened his mouth to respond—but stopped.

Because something about that question hit too close to home.

Hadn't he spent years assuming that intelligence was about logic? Hadn't he dismissed anything that couldn't be measured? Hadn't he ignored anything that did not fit within his framework of reality?

And if that was true...

Had he ever truly seen anything clearly?

The Intelligence That Must Be Lived

"You believe intelligence can be learned," the old man continued. *"That it is something you gather, like numbers on a page, like facts in a book."*

"Of course," Keshav said. *"We become more intelligent by learning."*

The old man chuckled. *"Then tell me—have you ever read a book about swimming?"*

Keshav blinked. *"Yes."*

"Good. Then tell me—could you swim just because you read about it?"

Keshav paused. *"No, of course not. You have to practice it."*

The old man nodded.

"Then why do you assume intelligence is any different?"

Keshav's mind stumbled.

"You believe that if you read enough, learn enough, study enough, you will understand intelligence," the old man said. *"But tell me—have you ever lived intelligence?"*

Keshav's breath caught slightly.

Because he didn't know how to answer that.

He had always assumed intelligence was about knowing things.

But what if real intelligence could only be understood through experience?

The First Shift in Perspective

"There is an intelligence that cannot be taught," the old man continued. *"One that cannot be measured, cannot be put into words, cannot be written in a textbook."*

"And what kind of intelligence is that?" Keshav asked cautiously.

The old man smiled.

"The intelligence that sees beyond facts. That perceives reality without distortion. That understands not with logic, but with clarity."

"And where does one find this intelligence?"

The old man's smile deepened.

"Not in books, not in theories, not in calculations."

He tapped his finger against Keshav's chest.

"In you."

Keshav tensed at the gesture.

Because he had spent his entire life looking outward for answers.

But what if the greatest truths were not out there, but within?

A New Kind of Knowing

Keshav was still skeptical.

"If this kind of intelligence exists, then why doesn't science acknowledge it?"

The old man chuckled.

"Because science measures what is external, what can be observed, what fits within its framework. But true intelligence is not an object—it is a state of being."

"And how do you measure a state of being?"

Keshav had no answer.

Because science had never been able to measure wisdom.

It had never been able to quantify insight.

It had never been able to explain why a person with no formal education could sometimes understand life better than a scholar.

Perhaps intelligence was not just about learning.

Perhaps it was also about unlearning.

About removing distortions, biases, and assumptions—so that one could truly see.

The Crumbling of a Belief System

The conversation had ended, but the shift had begun.

For the first time in his life, Keshav felt his rigid view of intelligence begin to break apart.

He had spent years chasing knowledge, believing it would lead to clarity.

But now, he saw the flaw in that thinking.

Because knowledge, without perception, was incomplete.

And intelligence, without wisdom, was blind.

A Step Toward the Unknown

As Keshav left the bookstore, the city around him felt different.

Not because it had changed—but because something within him had.

He wasn't sure where this path would lead him.

He wasn't sure if he was ready for what he might find.

But for the first time in his life, he was beginning to see.

And that was a kind of intelligence he had never known before.

A Conversation That Cannot Be Forgotten

Keshav watched as the old man walked away, his footsteps disappearing into the crowded street outside the bookstore.

He didn't say goodbye.

He didn't leave any final words of wisdom.

He simply left.

As if he had never been there at all.

As if the entire conversation had been nothing more than a ripple in time, a disturbance in the predictable rhythm of Keshav's mind.

And yet, that ripple refused to fade.

Keshav stood frozen in place, the weight of their exchange pressing down on him in a way no academic debate or intellectual argument ever had.

His scientific arrogance had been shaken.

His mind, once so rigid in its certainties, now felt like a landscape after an earthquake—cracked, uneven, unfamiliar.

He had walked into the bookstore in search of clarity.

Instead, he had found a question deeper than any he had ever asked before.

The Weight of a Question Without an Answer

Keshav had spent his entire life believing that every question had an answer, if one only searched hard enough.

That science, intelligence, and reason would eventually illuminate all mysteries.

But the old man had planted something in his mind that no formula could resolve.

"You think you know intelligence, but you only know its shadow."

The words refused to leave him.

Because he knew—deep down, though he hated to admit it—that the old man was right.

He had spent years studying the mechanics of intelligence—but had he ever truly understood what intelligence was?

Had he ever experienced it beyond logic, beyond calculation, beyond the cold precision of data?

Had he ever seen intelligence not as a function, but as a state of being?

The thought unsettled him.

Because it meant his understanding of intelligence, of knowledge itself, had been incomplete all along.

Walking in a World That No Longer Felt the Same

As he stepped out of the bookstore, the familiar noise of the city greeted him.

Cars honked. People moved past him, caught up in their own lives. Conversations drifted through the air, blending into an indistinct hum.

It was all the same.

Yet, he no longer felt like he was walking through the same world.

Something had changed.

Not outside of him, but within.

For the first time, he felt like he was seeing everything differently.

Not in the way he had been trained to—not as an observer, a scientist, a collector of data.

But as something else.

As if the filters of logic and calculation that had always shaped his reality had suddenly weakened.

And now, he was left staring into the unknown.

The Collapse of Certainty

Keshav had always relied on certainty.

On facts, evidence, reason.

He had built his world upon the idea that intelligence was the highest power, the ultimate force that would one day explain everything.

But today, he had glimpsed something beyond that.

Something that couldn't be calculated.

Something that couldn't be neatly categorized.

Something that whispered:

"There is more to intelligence than you know."

And that thought, more than anything else, terrified him.

Because it meant that all his life, he had been asking the wrong questions.

That there were answers that could never be found in books, in equations, in logic.

That true intelligence was not about knowing—but about seeing.

And now that he had glimpsed it, he could never go back.

A Restless Mind That Refuses to Let Go

That night, Keshav sat at his desk, staring at a blank page.

He had intended to write down his thoughts—to try and make sense of what had happened.

But no words came.

For the first time in his life, his rational mind was failing him.

The equations, the models, the intellectual frameworks he had always relied on felt useless in the face of what he had encountered.

He had spent his life believing that science was the path to truth.

But now, he wasn't sure anymore.

Because if science could not explain what he had felt today—

Then what else had it failed to explain?

The Beginning of a New Journey

Keshav had always believed that knowledge was the answer to everything.

But now, he wasn't even sure what the question was.

He had spent years chasing intelligence—believing that it would bring him clarity, fulfillment, purpose.

But now, he felt further from those things than ever before.

Because if intelligence alone was not enough...

Then what was?

He closed his eyes, exhaling slowly.

There was no going back now.

The old man had not given him an answer.

He had given him something far more dangerous.

A question that refused to be ignored.

And Keshav knew—this was only the beginning.

Part III: The Breaking Point

Choosing the Path of a Monk

"Science is no longer about truth. It's about funding."

8: The Collapse of the Illusion

The illusion didn't collapse in a single moment.

It didn't come crashing down like a sudden storm.

It cracked first—small, almost imperceptible. A feeling. A question. A realization that something wasn't quite right.

Then, the cracks spread.

Until, one day, everything he had built his life upon gave way beneath him.

And Keshav found himself standing in the ruins of his own certainty.

For weeks, he had been grappling with the idea that intelligence alone was not enough. That logic and reason, the foundations of his belief system, had limits.

But now, he wasn't questioning it in theory.

Now, he was facing it in reality.

Because the world he had placed his faith in—the world of science, logic, and progress—had betrayed him.

And no equation, no analysis, no calculated reasoning could fix what had just broken.

He sat in his office, staring at the screen in front of him, the words on the email blurring as his mind struggled to process what had just happened.

This wasn't an abstract debate anymore.

This was real.

Science wasn't saving him. Logic wasn't easing his pain. Intelligence wasn't giving him control.

For the first time in his life, he understood what it meant to feel truly lost.

And in that moment, the illusion shattered completely.

The Crisis That Shatters Everything

The email sat open on his screen, its words stark and final.

"Due to shifts in funding priorities, your research grant will be redirected to projects of higher commercial viability. Effective immediately, all ongoing work under this grant will be discontinued."

The words blurred, sharpened, then blurred again, as if his brain refused to fully process them.

For a moment, Keshav just stared, waiting for his mind to catch up with reality.

Everything he had built his career around, everything he had worked toward, was being discarded—not because it was flawed, but because it wasn't profitable enough.

Not valuable enough.

Not marketable enough.

It had nothing to do with truth, with scientific merit, with pushing the boundaries of knowledge.

It had everything to do with money, control, and power.

His hands clenched into fists as the realization hit him.

"Science is no longer about truth. It's about funding."

Joshi's words from weeks ago echoed in his mind.

At the time, he had resisted the idea. He had wanted to believe that science was still pure, that the pursuit of knowledge transcended corporate greed and political agendas.

But now, he saw the truth with brutal clarity.

The illusion had collapsed.

And he was standing in the wreckage.

A World That No Longer Made Sense

Keshav pushed his chair back from his desk, standing abruptly.

He needed air. Needed to escape the suffocating weight of betrayal.

He stepped outside, the cold night air hitting his face as he walked blindly through the empty streets near the research facility.

His mind raced, trying to make sense of what had just happened.

He had been working on something groundbreaking—a model that combined AI with complex astrophysical simulations, a project that could have revolutionized the way humanity understood deep space navigation.

But now?

Now, all of it was gone.

Because somewhere in a corporate boardroom, people who didn't even understand the work had decided it wasn't profitable enough to continue.

His research had been reduced to a line item in a budget report.

His years of work had been erased with a single email.

It didn't matter that he had dedicated his life to science. It didn't matter that he had spent sleepless nights chasing ideas, refining models, pushing his own limits for something greater. It didn't matter that this research could have changed the way humanity reached the stars.

In the end, it had no value—not because it wasn't important, but because it didn't serve the interests of those funding it.

And that truth was devastating.

Because for the first time in his life, Keshav felt powerless.

Intelligence Couldn't Save Him

Keshav had always believed that intelligence was the ultimate power.

That those who were smart enough, who understood enough, who mastered enough knowledge, could shape reality itself.

That science and logic could solve anything.

But now, he saw the limits of that belief.

Because intelligence had not protected him from this. Logic had not given him control over this. Science had not saved him from the forces that decided what mattered and what didn't.

For all his intelligence, he was powerless.

His mind, the very thing he had relied on for his entire life, could do nothing to change this reality.

And that realization shook him to his core.

The Cracks in the Foundation

He had spent his entire life believing in a rational world.

A world where truth prevailed over manipulation, where reason triumphed over chaos, where intelligence meant progress.

But if that world had ever existed, it did not exist anymore.

And now, he was standing at the edge of a different reality.

A world where science was a tool, not for truth, but for power. A world where intelligence, instead of bringing clarity, had only made him see how broken everything was. A world where those with knowledge were not the ones making decisions—the ones with money and influence were.

It was a world he had never wanted to believe in.

But now, he had no choice.

Because this world had just torn apart everything he had built.

What Had It All Been For?

Keshav kept walking, barely aware of where he was going.

His mind spiraled with a question he couldn't shake.

"What was all of this for?"

What was the purpose of knowledge if it could be discarded so easily? What was the value of intelligence if it couldn't protect itself from corruption? What was the meaning of science if it wasn't serving truth, but profit?

He had dedicated his life to something that no longer felt real.

And if science was no longer about seeking answers, about understanding reality, about expanding human knowledge—

Then what was it?

And more importantly...

Had he been chasing an illusion all along?

The Breaking Point

Keshav reached a park bench and sat down heavily, his hands trembling slightly.

The sky above was clear, vast, endless.

For years, he had looked at that sky and seen order.

Laws of physics governing motion, gravity, time. Equations that mapped the cosmos, that explained the dance of galaxies. Logic that told him, with absolute certainty, that the universe was predictable, structured, understandable.

But now?

Now, it felt chaotic.

Not because the universe had changed—but because his faith in how he understood it had crumbled.

For the first time, he realized—

The problem was not intelligence.

It was how intelligence was being used.

And if intelligence was no longer serving truth—

Then what was it serving?

A Truth He Could No Longer Ignore

As he sat in the cold silence, the full weight of his realization pressed down on him.

He had always believed that ignorance was the greatest enemy of humanity.

That the solution to the world's problems was more intelligence, more knowledge, more understanding.

But now, he saw the flaw in that belief.

Because evil did not come from ignorance.

It came from intelligence without balance.

The greatest threats to humanity had never been from the uneducated, the uninformed, the unaware.

The greatest threats had come from those who were brilliant—yet used that brilliance for power, control, destruction.

Wars were not started by fools. Exploitation was not engineered by the unthinking. Manipulation was not the tool of the ignorant.

It was the tool of those who knew exactly what they were doing.

And now, for the first time in his life, Keshav saw intelligence not as a solution, but as a weapon.

And if intelligence could be used as a weapon...

Then who was deciding how it was wielded?

Science, Career, and Logic Offer No Comfort

Keshav sat at his desk, staring at the blinking cursor on his screen.

The numbers in front of him should have been comforting.

Equations, algorithms, simulations—these were things that made sense.

For years, whenever life felt uncertain, he had always turned to his work. Data was reliable. Science was rational. Numbers never lied.

But today, none of it gave him peace.

He had spent the last few hours trying to force himself to focus, to lose himself in the familiarity of research, to drown out the storm inside his mind with the cold precision of logic.

Yet, no matter how many equations he ran, how many articles he skimmed, how many lines of code he typed...

The unease remained.

The doubts lingered.

Science wasn't saving him.

Logic wasn't giving him control.

And for the first time, he saw the limits of everything he had built his life around.

Searching for Stability in a Collapsing World

He had always believed that intelligence was the answer to everything.

That when faced with chaos, the best thing one could do was turn to knowledge, to rationality, to facts.

And yet, here he was, drowning in a kind of uncertainty that science could not fix.

His research—the work he had dedicated years to—now felt like a hollow exercise.

He could run as many calculations as he wanted, but they wouldn't tell him how to make sense of what he was feeling.

He could analyze data, but it wouldn't help him understand why everything felt so meaningless.

For the first time, he saw that there were problems intelligence could not solve.

The Failure of Pure Rationality

Keshav closed his laptop and leaned back in his chair, staring at the ceiling.

He had always dismissed emotions as secondary to logic.

A distraction. A complication. A primitive remnant of evolution that clouded rational decision-making.

But now, it wasn't logic that was failing him.

It was something else.

Something deeper.

He had always assumed that as long as he had knowledge, he would always have control.

But now, he had knowledge, and he felt more lost than ever.

And the most unsettling part?

He wasn't alone in this.

The Empty Lives of the Most Intelligent Minds

Keshav had spent his career working alongside some of the brightest minds in the world.

Physicists. Mathematicians. Engineers. AI researchers.

People who had unlocked the mysteries of the universe, built machines that defied human limitations, written codes that controlled the digital world.

And yet...

Many of them were deeply unhappy.

Some had driven themselves to exhaustion, chasing academic prestige that never fulfilled them. Some had isolated themselves from the world, unable to connect with anyone outside of their own logic-driven reality. Some had abandoned their ideals, selling their intelligence to corporations, governments, and industries that didn't care about truth—only about power.

He had once believed that intelligence led to understanding.

That the smarter someone was, the more clarity they would have.

But now, he saw that intelligence—without something to guide it—often led to emptiness.

To a world that was explainable, but not meaningful.

To answers, but not to peace.

The Limits of Science

For the first time, Keshav saw the invisible wall that science could never cross.

Science could tell him how the universe worked—but not why it existed. Science could explain the mechanics of love—but not the experience of it. Science could measure the brain's response to grief—but not the weight of loss.

It could describe suffering, but not heal it.

It could quantify emotions, but not capture their depth.

It could simulate intelligence, but not wisdom.

For years, he had believed that all truths could be reached through logic.

But what if some truths could only be understood through experience?

What if there were things that could never be explained—only felt?

The thought made his stomach twist.

Because it meant that everything he had built his life upon was incomplete.

The Realization He Didn't Want to Have

Keshav stood up and walked to the window, looking out at the city skyline.

For years, he had believed that knowledge was enough.

That intelligence, reason, and scientific inquiry were the highest forms of human achievement.

But now...

Now, he saw the flaw in that belief.

Because intelligence, on its own, had not saved him from this moment.

Logic had not given him answers.

And science, for all its brilliance, felt distant—cold and indifferent to the human experience.

He had spent his life looking outward.

Studying the laws of physics, the complexity of AI, the structures of intelligence.

But he had never once turned that intelligence inward.

Never questioned whether he had been looking in the wrong direction all along.

And now, for the first time...

He had to.

A Question That Science Couldn't Answer

Keshav closed his eyes.

The realization was painful.

Because it meant that intelligence alone wasn't the answer.

That the smartest people weren't necessarily the wisest.

That progress, without wisdom, was not progress at all.

That a world built purely on logic could never be truly complete.

He had always thought that science would give him clarity.

But now, he saw the truth—

Science could give him knowledge, but it could not give him meaning.

And that...

That was something he had never been ready to face.

The Moment of Surrender

Keshav sat back down, exhaling slowly.

His entire life had been about seeking answers.

But now, he was beginning to realize...

Some answers could not be found in equations.

Some questions had no logical solution.

Some truths could not be calculated, only lived.

And if that was true...

Then he had no choice but to do something he had never done before.

Something that terrified him.

He had to stop searching.

And start seeing.

The Realization: The Problem Is Not Intelligence—It's How We Use It

Keshav sat in his darkened office, the glow of his computer screen reflecting off his tired eyes.

For hours, he had stared at the blank page in front of him, his fingers hovering over the keyboard, waiting for his mind to organize itself into something rational.

But rationality had failed him.

Science had failed him.

Logic had given him no comfort.

He was at his lowest, disillusioned with everything he had once believed in.

And yet, it was in this moment—this deep, crushing silence—that something shifted.

A realization that should have been obvious.

A truth that had been staring at him all along, but one he had never truly seen.

"The problem is not intelligence."

"The problem is how we use it."

Intelligence Is Not the Enemy—But It Can Be Dangerous

For years, he had believed that ignorance was humanity's greatest flaw.

That if people were simply smarter, more rational, more knowledgeable, the world would be a better place.

But now, he saw how dangerous that assumption was.

Because intelligence alone didn't guarantee anything.

It didn't ensure goodness. It didn't ensure wisdom. It didn't ensure morality.

It only ensured capability.

And capability, without balance, was the very thing that had created the world's greatest horrors.

Intelligence had not saved the world from war—it had built more efficient weapons. Intelligence had not ended suffering—it had created systems of exploitation that ran with machine-like efficiency. Intelligence had not freed humanity—it had developed technologies that manipulated people without them even realizing it.

He had been blind to it before.

Because he had always seen intelligence as the highest good.

But now, he understood—

It was not intelligence itself that determined the future.

It was the hands that wielded it.

The Great Mistake of Human History

Keshav exhaled sharply, rubbing his temples.

He thought back to history—the moments that had defined civilization.

The greatest failures of humanity had not come from stupidity or ignorance.

They had come from brilliance, unchecked and unbalanced.

The greatest tragedies had not been caused by people who didn't understand the world.

They had been caused by people who understood it too well—and used that understanding for power.

The people who built nuclear bombs were not fools. The people who engineered financial collapses were not clueless. The people who created systems of mass surveillance were not unaware.

They were intelligent.

Highly intelligent.

But intelligence, without wisdom, without morality, without restraint, had led to destruction.

And now, Keshav saw the flaw in his belief system.

It wasn't ignorance that was the problem.

It was intelligence without balance.

Intelligence as a Tool—Not a Virtue

Keshav had always believed that intelligence was a virtue.

That those who possessed it were inherently more capable, more deserving of leadership, more suited to shape the future.

But now, he saw intelligence for what it truly was.

Not a virtue.

Not a moral compass.

Just a tool.

A tool that could be used to build or to destroy.

A tool that could be used to enlighten or to manipulate.

A tool that, when placed in the hands of the wrong people, became more dangerous than any weapon ever created.

And the terrifying part?

The wrong people had been wielding it for a long time.

How Intelligence Becomes Corrupt

Keshav thought back to his work—to the research projects that had been shut down, to the scientists who had been forced to mold their findings to fit corporate and political interests.

Intelligence, in theory, was pure.

But in the real world, it was never free from influence.

It was controlled by those who had the money, the power, the connections.

It was used not to seek truth, but to serve those in control.

And he had been naïve to think otherwise.

He had believed that science was beyond corruption.

That knowledge was always leading humanity forward.

But now, he saw the reality.

Progress was not guaranteed.

Because progress was determined by who controlled intelligence—and what they used it for.

And history had shown that power rarely served truth.

The Truth He Could No Longer Ignore

Keshav looked around his office, at the books, the research papers, the whiteboards filled with formulas.

For years, he had thought that these things held all the answers.

But now, he understood—

Knowledge is not power.

Wisdom is power.

Because knowledge, in the wrong hands, became manipulation.

Intelligence, without balance, became destruction.

And science, without wisdom, became just another tool of control.

He had spent his life worshiping intelligence.

But now, he realized—intelligence alone was not enough.

And that realization changed everything.

A New Path—But No Clear Direction

Keshav closed his laptop.

For the first time, he wasn't searching for an answer.

Because he knew that the answer couldn't be found in data, in equations, in logic.

It had to be lived, not calculated.

He didn't know what came next.

He didn't know what he would do.

All he knew was that he couldn't keep following the same path.

Because he could no longer pretend that intelligence was the answer.

It was how intelligence was used that mattered.

And for the first time in his life, he wasn't sure if humanity was using it wisely.

Evil Is Not Ignorance—It Is Intelligence Without Balance

For most of his life, Keshav had believed that ignorance was the root of all evil.

That war, corruption, destruction—all of it stemmed from a lack of understanding, from the failure to think rationally.

That if only people were more intelligent, more knowledgeable, more logical, the world would become a better place.

But now, he saw how dangerously naïve that belief had been.

Because history told a different story.

A story where evil was not the result of stupidity—but of intelligence, twisted and unchecked.

A story where some of the most brilliant minds had caused the greatest harm.

And for the first time, Keshav understood—it was not the absence of intelligence that led to destruction.

It was intelligence without balance.

The Danger of Purely Rational Thinking

Keshav leaned back in his chair, his fingers tracing circles on the desk as his thoughts unraveled before him.

Logic, by itself, had no morality.

It simply calculated outcomes.

It was cold, detached, indifferent.

It didn't care if those outcomes were just or unjust, ethical or unethical.

It only cared about efficiency, about results, about achieving the most optimal path toward a goal.

And therein lay the danger.

Because without wisdom, without ethics, without restraint, intelligence could justify anything.

If an empire wanted more resources, logic could justify war. If a corporation wanted more profit, logic could justify exploitation. If a government wanted more control, logic could justify oppression.

And it had.

Again and again, throughout history, intelligence had enabled destruction as much as it had enabled progress.

It had not always been used to enlighten.

Sometimes, it had been used to enslave.

History's Greatest Crimes Were Not Committed by the Ignorant

Keshav thought back to the greatest tragedies in human history.

The ones that had shaped nations, that had altered the course of civilization.

And he saw something he had never noticed before.

They were not caused by ignorant fools acting on impulse.

They were caused by highly intelligent individuals who knew exactly what they were doing.

The architects of genocides were not mindless thugs. They were strategists, policymakers, men in suits who used intelligence to plan mass destruction.

The corporations that manipulated economies, driving millions into poverty, were not clueless. They were filled with brilliant economists, financial experts, and data analysts who designed the perfect systems of exploitation.

The regimes that controlled information, twisting narratives to shape entire populations, were not led by the uninformed.

They were engineered by the best minds in propaganda, psychology, and technology.

Time and time again, it was not ignorance that caused the greatest suffering.

It was the precise, calculated application of intelligence—without morality, without restraint, without wisdom.

And that realization sent a chill through him.

Because it meant that intelligence alone did not prevent evil.

If anything, it made it more efficient.

The Rationalization of Atrocities

Keshav had always believed that rationality was a safeguard against wrongdoing.

That if people thought critically, reasoned logically, analyzed things objectively, they would naturally make better decisions.

But now, he saw how flawed that assumption was.

Because rationality, without morality, could justify anything.

History had shown that the worst atrocities had always been rationalized.

They had never been committed out of blind rage, but out of careful calculation.

Genocides had been justified as "necessary for national security." Economic exploitation had been justified as "necessary for growth." Human suffering had been justified as "an unavoidable consequence of progress."

Intelligence, when disconnected from empathy, had created excuses instead of ethics.

And now, Keshav saw the terrifying truth—

Logic was only as good as the hands that used it.

And if those hands lacked wisdom, lacked morality, lacked humanity...

Then intelligence was not a force for good.

It was a weapon.

The Myth of the "Good Scientist"

For years, Keshav had believed that scientists were different.

That those who pursued knowledge were inherently above greed, power, and corruption.

But now, he wasn't so sure.

Because history had shown that brilliant scientists had often enabled the worst horrors.

The atomic bomb was not created by warlords. It was created by physicists.

Mass surveillance was not developed by dictators. It was designed by engineers.

Weapons of war, chemical agents, AI-driven manipulation— None of it came from the minds of the ignorant.

It came from the minds of the highly intelligent.

People like him.

And that terrified him.

Because it meant that intelligence, without ethics, did not bring salvation.

It brought efficiency to destruction.

And if intelligence wasn't guided by something greater than logic...

Then who was stopping it from turning against humanity itself?

The Missing Piece: Intelligence Without Wisdom Is Blind

Keshav rubbed his temples, his mind unraveling at the seams.

He had spent his life believing in intelligence as the highest pursuit.

But now, he saw that intelligence wasn't enough.

That intelligence, without balance, without wisdom, without restraint, was just another form of power.

And power, when left unchecked, was never neutral.

Power, when given to the wrong hands, became corruption.

And now he understood—

The missing piece had never been intelligence itself.

It had been how intelligence was used.

It had been who controlled it.

It had been what guided it.

And for the first time, Keshav saw intelligence not as a solution, but as a responsibility.

Because without wisdom, intelligence was blind.

And without morality, it was dangerous.

And now, for the first time, he understood—

The world's greatest problems were not caused by ignorance.

They were caused by intelligence—misused, unchecked, and without balance.

A New Fear—And a New Question

Keshav stood up, walking toward the window.

The city lights flickered below him, a world built on intelligence, on science, on progress.

But progress wasn't guaranteed.

It was decided.

By who controlled intelligence.

By who dictated how it was used.

And by who had the wisdom—or the lack of it—to shape the future.

For years, Keshav had believed that intelligence would save the world.

Now, he wasn't sure if it would destroy it first.

And for the first time, a new fear crept into his mind.

A fear more terrifying than ignorance itself.

If intelligence, without balance, could become the greatest force of destruction...

Then who, if anyone, could be trusted to wield it?

The Problem Is Not Faith—It's That People Follow It Blindly

Keshav had never given faith much thought.

To him, faith was the opposite of intelligence.

It was belief without evidence, trust without proof, surrender without questioning.

It was what people turned to when they didn't have answers.

And for someone like him—someone who had spent his life searching for truth through logic, through reason, through facts—faith had always seemed like a weakness.

Something for those who couldn't think for themselves.

But now, as he sat alone in his office, wrestling with the disillusionment of science, intelligence, and rationality, a strange thought entered his mind.

What if faith wasn't the problem?

What if the problem was how people used it?

What if faith, like intelligence, was simply a tool?

And what if the real danger was not in having faith, but in following it blindly?

Faith and Intelligence: Two Sides of the Same Coin

For years, Keshav had dismissed faith as a crutch for the weak-minded.

A way for people to avoid thinking, to escape the responsibility of questioning.

But now, as he looked at the world through the lens of everything he had recently realized, he saw something disturbing.

Faith and intelligence were not so different after all.

Both could be used for good. Both could be used for harm. Both could bring clarity—or lead to destruction.

Faith, when guided by wisdom, could bring purpose, balance, and meaning.

But faith, when blind, could be just as dangerous as blind intelligence.

Because blind faith led to dogma. Blind faith led to wars fought in the name of gods. Blind faith led to people following orders without questioning.

And wasn't that the same thing that had made intelligence dangerous?

When intelligence was guided by wisdom, it advanced humanity.

But when intelligence was followed blindly—when people surrendered to logic without morality, when they justified destruction in the name of progress—it became a weapon.

Faith worked the same way.

The problem wasn't faith.

The problem was people surrendering to it without thinking.

And the more Keshav thought about it, the more he realized—

Blind faith and blind intelligence were two sides of the same coin.

Both could justify anything.

Both could lead to destruction.

Faith Without Questioning vs. Faith With Wisdom

Keshav leaned back in his chair, staring at the ceiling.

What made faith different from intelligence wasn't that one was good and the other was bad.

It was that both needed something to guide them.

Faith, when combined with wisdom, could create deep purpose.

It could help people find balance in a chaotic world.

It could bring peace, resilience, and a sense of connection to something greater.

But when faith was blind, unquestioning, unchallenged?

It became a tool for control.

A way for people to stop thinking, to follow orders, to justify actions without accountability.

That was why people could be manipulated by leaders, by religious institutions, by political ideologies—because they surrendered their minds to something without ever asking if it was right.

And wasn't that exactly what had happened in science and intelligence too?

Brilliant minds had followed governments, corporations, institutions—never questioning whether their research was being used for good or evil.

Just like blind faith, blind intelligence had caused its own disasters.

And now, Keshav saw that humanity's greatest problem had never been intelligence vs. faith.

It had been blindness.

The failure to question. The failure to think for oneself. The failure to see beyond what was given.

And that realization shook him.

Because it meant that he had been blind, too.

The Danger of Surrendering Without Thinking

For years, Keshav had prided himself on being logical, rational, and scientific.

He had always thought that he was different from those who followed faith.

That he was smarter, more enlightened, more capable of seeing the truth.

But now, he saw the flaw in his thinking.

Because he had followed science with the same kind of blind faith that religious people followed their gods.

He had assumed that science was always right. That progress always meant good. That intelligence was always the answer.

He had never stopped to ask, "What if I'm wrong?"

What if intelligence, without wisdom, was just as dangerous as faith without questioning?

What if science, when unchecked, became its own kind of religion?

A belief system where people followed data without considering the human cost.

A system where facts were worshipped, but meaning was ignored.

A world where logic ruled, but morality was an afterthought.

And wasn't that exactly what had happened?

The Search for Balance

Keshav exhaled slowly.

He had spent his whole life rejecting faith.

But now, he understood why people needed it.

Not because they were weak.

Not because they were afraid.

But because faith, when guided by wisdom, gave people something intelligence never could—meaning.

And meaning wasn't found in equations.

It wasn't found in logic.

It wasn't even found in intelligence.

It was found in how one chose to live.

And now, Keshav understood—

The answer wasn't intelligence or faith.

The answer was balance.

A world where people didn't follow blindly, but also didn't reject things they didn't yet understand.

A world where science and wisdom worked together, instead of fighting against each other.

A world where intelligence didn't rule without ethics—and faith didn't exist without reason.

And for the first time in his life, he saw a new path.

One that wasn't about choosing between intelligence or faith—

But about understanding both, questioning both, and finding the wisdom that lay in between.

The Shift in Perspective

Keshav closed his eyes.

Everything he had believed in was changing.

His faith in intelligence had collapsed.

His rejection of faith had been challenged.

And now, he stood at a crossroad.

One path led back to where he had been—to the safety of logic, science, and reason.

The other led somewhere unknown—toward something deeper, something beyond intelligence alone.

And though he didn't know where it would take him, he knew one thing for certain.

He could no longer follow anything blindly.

Not science.

Not faith.

Not even himself.

For the first time, he wasn't searching for certainty.

He was searching for truth.

And truth, he now realized, could only be seen by those willing to question everything.

Even the things they once believed were unshakable.

Even the things they had built their entire lives upon.

And now, Keshav was ready to do just that.

A New Path Begins to Emerge

Keshav had spent his life looking outward.

Toward science, technology, intelligence.

Toward the laws of the universe, the mechanics of reality, the precision of logic.

He had always believed that truth lay in data, in research, in things that could be measured and analyzed.

But now, for the first time, he saw that he had been searching in only one direction.

That the answers he had been looking for did not exist in equations alone.

That intelligence without wisdom was incomplete.

That logic without perception was blind.

That knowledge without experience was hollow.

And for the first time, Keshav saw a different path.

One that was not about choosing between intelligence and faith, between science and meaning, between reason and experience.

One that was about balance.

The Search for Something More

Keshav sat in his office, staring at the blank page on his laptop.

He had spent weeks trying to return to the life he had known.

To immerse himself in science, to lose himself in work, to drown out his doubts with calculations and research.

But it wasn't working.

Because he could no longer ignore what he had seen.

The realization that intelligence alone was not enough. That knowledge, without wisdom, could be dangerous. That logic, without understanding, was cold and empty.

He had spent so much time thinking, analyzing, proving.

But had he ever truly understood?

Had he ever lived intelligence rather than studied it?

Had he ever trusted what he saw rather than what could be calculated?

And now, he knew—

If he was going to move forward, he had to change the way he searched.

The Difference Between Knowledge and Wisdom

For years, Keshav had dismissed philosophy, intuition, and introspection.

To him, these were imprecise, unscientific, subjective.

But now, he saw that they were not separate from intelligence.

They were part of it.

Because knowledge could be learned, but wisdom had to be lived.

Because intelligence could calculate, but it could not always perceive.

Because the greatest minds in history had not only been thinkers—they had been seekers.

They had not only pursued data—they had pursued truth.

And truth, he now realized, was more than just facts.

It was how one saw the world.

How one chose to use what they knew.

How one lived.

And for the first time, Keshav saw that his search for truth had to be more than intellectual.

It had to be experiential.

The End of the Crisis—The Beginning of Something Deeper

He had once seen this as a crisis.

A collapse of everything he had believed in.

A shattering of the world he had built around logic, intelligence, and science.

But now, he saw it for what it really was.

Not an end.

A beginning.

A doorway into a new way of seeing.

One where intelligence was not just knowing—but perceiving. One where logic was not just thinking—but understanding. One where truth was not just external—but internal.

For the first time, he did not feel the need to prove anything.

For the first time, he did not feel the urge to rationalize everything.

For the first time, he did not have all the answers.

But that was okay.

Because for the first time—he knew what he was searching for.

And this was no longer just about intelligence.

This was about something deeper.

Something he couldn't yet define.

But something he was finally ready to find.

9: The True Nature of God

The Ultimate Realization

The plane touched down in Mumbai just after dawn.

Keshav looked out of the window—not with the usual detachment of a man returning to a familiar country, but with the quiet anticipation of someone arriving for the first time.

India had always been his origin. But this time, it felt like his destination.

He had spent his life building equations and theories that reached for the stars. But now, something deeper called him back to the ground beneath his own feet.

Not to prove anything. Not to teach or to argue. But to listen.

For the first time, his mind was not searching for a formula—it was searching for understanding.

And he knew—whatever came next could not be found in textbooks or telescopes.

It had to be found in something older, something forgotten, something sacred.

He had reached the end of one kind of knowledge. And at that edge, something else was waiting.

A truth whispered not in numbers, but in silence. A presence not described by science, but felt by the soul.

And in that quiet knowing, Keshav took his first step—not as a scientist. But as a seeker.

Returning to the Roots

The air was thick with the scent of incense and old monsoon dust. The streets of India moved with a rhythm that Keshav had once taken for granted—chaotic yet strangely synchronized, wild yet intimate, always pulsing with something ancient.

But this time, he wasn't here for nostalgia. He wasn't here as a visiting scientist or a celebrated name in the global research circuit.

He was here for something deeper—something he could no longer name through logic alone.

He had returned not as a man of intellect, but as a man who had finally realized the limits of intellect.

Not a Vacation. A Pilgrimage.

From the moment he stepped off the plane, Keshav felt the shift—not in the world around him, but in himself.

Everywhere he went—whether through crowded alleys in Banaras, the echoing halls of ancient temples in Tamil Nadu, or the quiet, sunlit verandas of Himalayan ashrams—he moved like a stranger in a world he was born into.

But perhaps that was necessary.

Perhaps to truly learn, one had to first forget what they knew.

And Keshav had spent a lifetime knowing. Now, he was ready to unknow.

He did not introduce himself as a scientist anymore. He simply asked questions.

He visited temples not for prayer, but for presence. He sat beside monks who didn't speak unless spoken to. He listened to wandering sadhus who quoted the Gita and quantum mechanics in the same breath.

He read translations of Upanishads at night and scribbled questions in the margins—not to dissect them like formulas, but to taste their essence.

He wasn't searching for God in the form of a deity.

He was searching for something far more difficult to grasp—the nature of intelligence itself. And its connection to everything.

Conversations, Not Lectures

In Rishikesh, he met a priest who had never gone beyond high school but spoke of consciousness as though he had lived lifetimes inside it.

Keshav asked him once, "Do you believe in God?"

The priest chuckled. "Believe? No. I try to understand."

Keshav raised an eyebrow. "Isn't belief the foundation of faith?"

The old man smiled. "Only when one lacks the courage to ask. But when you ask deeply enough... belief turns into direct experience. That is real faith. That is understanding."

And that hit Keshav like a silent thunder.

He realized that in his entire scientific career, he had prided himself on questioning everything—except the faith of others.

He had judged it from the outside, not realizing that many of them were not blindly following myths.

They were living realities that science hadn't even begun to understand.

The Shift: From Analysis to Absorption

Keshav had always relied on the mind to process truth.

But now, as he sat beside the Ganga in silence, watching the water flow with timeless grace, he began to feel something he could not explain.

Not awe. Not reverence. Something else.

A sense that intelligence was not something humans possessed—It was something we existed within.

Something that expressed itself through us, just as it did through rivers, trees, stars, and silence.

And for the first time, he didn't try to reduce that feeling into terms. He didn't try to define it with precision. He simply watched it.

Felt it.

And in doing so, something shifted. Something opened.

He no longer wanted to understand the Divine. He simply wanted to experience it.

Why Wisdom Was Missing from His Intelligence

Everywhere he went—whether speaking to a blind poet in Gujarat or a retired yogi in Kerala—Keshav encountered a strange consistency:

These people did not seek to impress. They did not speak in citations or boast about discoveries.

And yet, their words carried more weight than most conferences he had ever attended.

Because they spoke from experience. Not knowledge—but wisdom.

And Keshav realized the great flaw in his own journey:

He had built towers of intelligence on a foundation that lacked inner silence. He had collected knowledge like trophies, but never allowed that knowledge to turn inward and become understanding.

He had known about the universe. Now, he wanted to know why it existed. Why *he* existed.

And that meant not just learning—but unlearning.

Not just hearing—but listening.

Not just thinking—but feeling.

Science Had Explained the 'How.' But Not the 'Why.'

In the ashram of a reclusive thinker in the Nilgiris, Keshav finally asked the question that had brought him across the world:

"If there is a force behind everything... if what you call God is real... what is it, really?"

The teacher, who had been silent all morning, looked up and said,

"God is not a being. God is intelligence manifesting with intent."

Keshav's breath caught.

He didn't understand it fully—yet something about that line landed deeper than any definition he had ever encountered.

And for the first time, he didn't rush to question it.

He simply let it echo.

A Journey Inward

India was no longer just a place to visit.

It had become a mirror.

Not of culture or heritage—but of something Keshav had long forgotten:

The Self.

Not the self made of achievements or thoughts. Not the self crafted by identity or profession.

But the silent self—the one that observed, that questioned, that listened.

The one that didn't need evidence to feel.

The one that already knew—long before it was taught.

And now, for the first time, Keshav wasn't resisting that self.

He was walking toward it.

And though he still didn't have all the answers— He finally understood what he was meant to seek.

Not success. Not certainty. Not even peace.

But the truth behind truth.

The intelligence behind intelligence.

The God behind the word "God."

And that journey, he knew, had only just begun.

The First Shock: God Is Not a Being

The first time Keshav read the phrase, he dismissed it as poetic.

"God is not a person. God is intelligence manifesting with intent."

It was buried in a footnote of a translated Upanishadic commentary, scrawled in faded ink by a monk whose name had long since disappeared into the folds of time.

He read it again.

And again.

Something about the line refused to fade.

Reading the Ancients with New Eyes

Keshav had spent his life viewing ancient scriptures as mythology cloaked in ritualism—stories told to comfort the primitive mind before science arrived.

But now, stripped of arrogance and awakened by crisis, he returned to these texts not with the mind of a scientist—but with the heart of a seeker.

And what he found stunned him.

The language was metaphorical, yes. The stories often symbolic. But the core ideas—they weren't irrational.

They were encoded truths—concepts that pointed to something deeper than description, beyond logic.

And over and over again, across Upanishads, Vedic hymns, Shaivite commentaries, and even Buddhist logic, one idea kept returning:

God is not a being. God is not a king in the sky. God is not a rewarder or punisher. God is intelligence. Not static. But in motion. Intelligence... with intent.

And for the first time, Keshav stopped reading to understand. He started reading to see.

Not a Creator Sitting Outside Creation

The idea was radical—especially to someone raised in the sterile corridors of astrophysics.

Science had always dismissed "God" as an external, unverifiable hypothesis. A creator God—separate from creation—was incompatible with scientific models.

But what if God was not separate? What if God was not outside the system, but the very intelligence that animated it from within?

What if God was not an entity to believe in, but a presence to recognize—like gravity, but deeper... intentional?

That realization shocked him.

It was the opposite of everything he had argued in debates, dismissed in lectures, and rolled his eyes at when others tried to explain.

God wasn't a person. God wasn't a figure. God wasn't something to pray to in exchange for blessings.

God was intelligence, choosing to move. Choosing to become.

It wasn't supernatural. It was ultra-natural.

And once he saw it, he couldn't unsee it.

God as Action, Not Just Awareness

In the Chandogya Upanishad, he found a phrase that echoed in his mind for days:

"Tat Tvam Asi" — That Thou Art.

It didn't speak of a distant God. It declared that you are part of that intelligence. That the same force that moves stars and atoms moves you.

God was not a passive backdrop. God was the impulse within. The intent to act. The will to transform. The intelligence behind manifestation.

And suddenly, Keshav began to reinterpret everything:

- The Big Bang? Not randomness—but intent embedded in energy.
- Evolution? Not blind mutation—but intelligence adapting itself through form.
- Human emotion? Not weakness—but intelligence

navigating complexity.

Everything he had once seen as disconnected, chaotic, or mechanistic now glimmered with intelligence expressing itself with purpose.

He was not separate from this process. He was an instance of it.

Seeing Through the Illusion of Divinity as a Figure

It became clear why so many had misunderstood divinity.

Humans had always tried to personalize the incomprehensible. They turned intelligence into a figure. Intent into personality. Energy into deities.

And while the stories had power, they had also created confusion.

We began to pray to names—Brahma, Vishnu, Shiva, Allah, Jesus, Buddha— but we forgot what those names pointed to.

Not people. Not forms. But manifestations of intelligence in different modes.

Keshav understood now:

The sages weren't idol worshippers. They were symbol architects.

They had encoded formless truths into relatable metaphors.

But as time passed, the metaphors became mistaken for the reality. And now, people debated whose God was true—while the truth itself waited patiently to be seen.

Intelligence with Intent: The Living God

Keshav returned to the phrase again: *"God is intelligence manifesting with intent."*

It was not blind energy. It was not random motion.

It was intelligence that chose to become. That observed and acted.

It had the power to create galaxies. And the power to weep through human eyes.

It was what made a seed grow into a tree. It was what made a mind wonder why.

And that changed everything.

Because this God—this intelligence in motion—was not outside the universe. It was the universe.

It was the mind behind evolution. The heart behind morality. The will behind life.

No More Belief. Only Recognition.

Keshav no longer needed to "believe" in God.

Because God was not something to believe in. God was something to recognize.

Like seeing light after being blindfolded. Like feeling warmth without needing to measure its temperature.

This wasn't surrender to religion. This was integration.

The God he had rejected as fiction... was simply a truth he had never been trained to perceive.

The Shock That Rewired Everything

Keshav closed the book, his hands trembling.

This wasn't a scientific breakthrough. This was an existential one.

All his life, he had searched for the "Theory of Everything."

But now, he saw that the theory wasn't just equations and particles.

It was intelligence itself.

Not just observing—but acting. Not just existing—but choosing.

And that... that was divinity.

Not the God of temples or textbooks. But the God that moved through both.

The shock was not that God existed. The shock was that God had always been right there—hidden in plain intelligence.

Waiting to be seen. Not believed.

Kindness and Cruelty: The Duality of Action

Keshav stood at the edge of the Ganga, the cool water lapping gently against his bare feet. Pilgrims chanted hymns behind him, bells echoed from distant temples, and incense swirled in the thick morning air. But his mind was anything but still.

If God—this force, this intelligence—was truly real, then why did the world seem so broken?

Why did children suffer? Why were good people silenced, and tyrants celebrated? Why did cruelty not only exist—but often thrive?

He had finally accepted that intelligence lay at the root of everything. But that only deepened the paradox.

If intelligence governed the universe, why did it feel so... unjust?

The Moral Dilemma That Science Never Solved

In science, suffering was biological.

A result of entropy, survival, and evolutionary error margins. Pain served a function—alerting organisms to danger or failure. Cruelty was framed in terms of behavioral science, hormones, tribal instincts, or power dynamics.

But now that he had accepted a deeper layer—intelligence with intent—Keshav found himself morally disturbed.

Because suddenly, suffering wasn't just a random by-product of nature.

It was part of a system governed by intelligence.

And that raised a terrifying question: Was intelligence... complicit? Or worse... indifferent?

He looked into the river, hoping for stillness. But the ripples only magnified his storm.

The Insight That Shifted Everything

Later that evening, while meditating at a hilltop monastery, an old monk sat beside him. They exchanged no greetings. Just silence.

After a long pause, the monk whispered, almost as if responding to Keshav's inner thoughts:

"When intelligence enters reality, it must act. And all action creates dual consequences."

Keshav opened his eyes. The monk was gazing at the sky.

"You want a world of light with no shadows. A universe of creation without destruction. But action... action always disturbs balance. It creates direction, motion, polarity."

And in that moment, the insight crystallized:

Intelligence is neutral until it chooses. But the moment it acts—it splits reality into consequences.

Every expression of intelligence was a disturbance. A ripple. A wave. And every wave, by nature, had a crest and a trough.

That was the cost of manifestation.

Creation and Destruction – Two Hands of the Same Force

Keshav recalled the concept of Shiva—the destroyer—not as an evil force, but as a necessary arm of cosmic balance.

He began to see that creation and destruction, kindness and cruelty, were not separate forces at war—but two ends of the same intelligent movement.

Creation cannot exist without destruction. Compassion cannot be understood without cruelty. Order cannot emerge without chaos resisting it.

When intelligence chooses to act, it enters duality—and duality creates consequence.

Not because the intelligence is evil. But because the very nature of existence is contrast.

The Problem Isn't That Evil Exists—It's That We Only Expect Good

It hit him hard.

All his life, Keshav had carried an unspoken assumption: If there is intelligence in the universe, then it must ensure happiness, justice, and comfort.

But why?

Was that expectation even logical? Or was it a human desire projected onto a cosmic truth?

He realized:

We do not suffer because intelligence failed. We suffer because we expect intelligence to only create comfort, not consequence.

Yet, every scientific principle he had ever studied showed the same law:

Action creates reaction. Force creates resistance. Choice creates polarity.

So how could divine intelligence escape this rule?

The answer was: it didn't. It expressed itself through it.

Cruelty and Kindness Are Both Faces of Intelligent Action

Cruelty did not mean the universe was evil. Kindness did not mean it was good.

Both were results of intelligence entering space and time, manifesting through circumstance, complexity, and consequence.

What people called "God's will" was not a plan of reward and punishment. It was simply the expression of intelligent force navigating reality—through human choices, natural systems, and karmic echoes.

A tree grows—because intelligence flows through its cells. A child starves—because other intelligent beings made choices to hoard resources. A disaster occurs—not as punishment, but as rebalancing in a larger system we barely comprehend.

Intelligence doesn't micromanage morality. It empowers action. And action creates both beauty and suffering.

That was not divine betrayal. It was divine honesty.

Understanding Doesn't Erase Suffering—But It Makes It Bearable

Keshav didn't feel comforted. Not in the conventional sense. But he did feel clarity.

The world still hurt.

There was still injustice. Still war. Still pain.

But now he saw these not as proof against God, but as the shadow of intelligence acting in an imperfect world.

He saw cruelty as the imbalance of intelligent choice. He saw kindness as the harmony of aligned intent.

And he saw that both were part of the same unfolding process. Not a test. Not a punishment. But a result.

And in recognizing that, something inside him settled. Not into peace. But into understanding.

Beyond Good and Evil: Toward Balance

The problem wasn't that cruelty existed. The problem was that humanity had not yet learned how to use intelligence wisely.

Just as a blade can harvest or kill, intelligence could create or corrupt.

When used with awareness, intelligence heals. When used without balance, it destroys.

Keshav realized the next step in his journey was not to escape suffering, but to understand how to align his intelligence with balance.

With intent that nurtures, not dominates. With action that integrates, not divides.

And that journey, he knew now, was the path toward something sacred. Not outside him. But through him.

Science and Faith Are Not Opposites

The sun was setting behind the temple domes of Varanasi, washing the Ganga in golden hues as Keshav sat on the worn steps, his notebook beside him, untouched.

He had spent the past weeks immersed in ancient scriptures and spiritual dialogues, reconciling the metaphysical with the measurable.

At first, he had kept his disciplines separate. Science was structure. Faith was story.

One dealt in data. The other in devotion.

He had spent his entire life believing they were incompatible. That one explained reality, and the other escaped from it.

But now—after everything he had seen, questioned, and felt—he realized:

Science and faith were never enemies. They were just misunderstood perspectives of the same truth.

Science Studies the Structure. Faith Seeks the Source.

In the monasteries of the south, he had studied Vedic physics—concepts of akasha (space), kala (time), and prana (energy) that paralleled quantum ideas with astonishing precision.

He had met monks who spoke of the atom as a vibration long before it was observed under electron microscopes. He had found hymns that echoed modern laws of thermodynamics.

And yet, none of them sought patents. None tried to claim scientific discovery. Because their goal was different.

They were not trying to explain how reality worked. They were trying to understand why it exists at all.

Science, he realized, was like dissecting a flower to study its anatomy. Faith was like smelling its fragrance to understand its essence.

Science studies the mechanism. Faith seeks the motive.

They Weren't Speaking Different Languages—They Were Answering Different Questions

All his life, Keshav had tried to solve the universe by asking *"How?"*

- How do stars form?
- How do neurons fire?
- How do algorithms mimic thought?

But now, the question that haunted him was: *"Why?"*

- Why do we seek purpose?
- Why does kindness matter in a mechanical world?
- Why does intelligence produce both creation and conflict?

Science gave him clarity. Faith gave him depth.

One satisfied the mind. The other awakened the spirit.

And suddenly he saw: neither was enough alone. Because structure without source is hollow. And belief without understanding is blind.

The Illusion of Conflict

He remembered how often he'd dismissed religious thinkers as irrational. And how often spiritual people had rejected science as arrogant.

But now he saw the tragedy— They had been arguing from different vantage points, not different truths.

It was like two people looking at the same mountain— one from the base, measuring its terrain, the other from the summit, feeling its vastness.

Both were right. And both were incomplete.

Science explains the visible. Faith explores the invisible.

But intelligence unites them.

It is the lens through which both gain clarity.

The Missing Third Eye—Intelligence

Sitting in a quiet temple courtyard, Keshav penned a thought in his journal:

"Science and faith are like two eyes—each sees a part of reality. But only together, through the lens of awakened intelligence, can they reveal the whole."

This was his realization. Not belief. Not theory. But integration.

Science was never meant to disprove faith. Faith was never meant to replace science.

They were both tools. Both limbs of the same seeker.

But they needed balance. They needed intelligence—not as IQ, but as insight. As the conscious thread that asks:

What am I seeing? And why?

The Sages Were Scientists. The Scientists Were Seekers.

As he dove deeper into the works of ancient Indian, Greek, and even Islamic scholars, Keshav saw something astonishing:

The sages of the past never separated the search for truth from the search for self.

Aryabhata studied celestial motions to understand the rhythm of existence. Patanjali taught breath not only for health, but for clarity of mind. The Buddha didn't claim divine revelation—he claimed awakened observation.

They were scientists of experience. Seekers of symmetry—between the outer universe and the inner one.

And somewhere along the way, modern thinking had fractured that unity.

Science became arrogant without introspection. Faith became hollow without understanding.

But now, in this collapsing divide, Keshav saw what had been lost:

Wholeness.

Faith Is Not the Enemy of Reason. It Is the Companion of Humility.

Faith was not the refusal to question. It was the courage to admit:

"There are truths beyond what I currently understand."

Faith wasn't surrendering thought. It was surrendering ego.

And science, at its best, also required faith— in pattern, in logic, in laws that no one could fully explain but everyone relied upon.

The difference wasn't in method. The difference was in intent.

Science seeks to control. Faith seeks to connect.

And true intelligence? It sees both. It uses both. It transcends both.

The Full Picture

Keshav finally understood that the world had been limping on one leg.

People chose science or faith, logic or belief, data or devotion—as if they were contradictions. But in doing so, they missed the third perspective—the integrative intelligence that makes both meaningful.

He no longer felt compelled to choose.

He could admire the elegance of physics and sit in silence under temple bells.

He could trust neural networks and still bow to the intelligence that flows through breath, birth, and being.

Because the world didn't need more debates.

It needed synthesis.

And that was the path he would now walk. Not as a scientist. Not as a spiritualist. But as a witness to both.

As someone who had seen that:

Science and faith are not opposites. They are complementary currents in the same ocean of truth.

Karma, Emotion, and Free Will—All Are Intelligence in Motion

By now, Keshav had shed the arrogance of answers and embraced the humility of questions. He no longer sought to master reality but to listen to it.

One afternoon, while walking barefoot along the muddy path of a rural village near Bodh Gaya, he watched a child accidentally drop a clay pot. It shattered, scattering water and shards in all directions.

The mother, startled by the sound, scolded him. The boy wept. But then, the father quietly walked over and began sweeping the mess—without anger, without blame.

Keshav stood frozen.

A single thought echoed in his mind:

"This... is karma in motion."

Not in the mystical, moralistic way he had always heard people use the word. But in the most practical, real, and observable sense.

The child acted. The pot broke. The consequence rippled outward—emotions, responses, choices. But nowhere was there judgment.

Only motion.

Karma: Not Punishment, But Pattern

Growing up, Keshav had dismissed karma as religious bookkeeping. A tool used to control people's behavior by threatening cosmic consequences. A superstition.

But now, standing in the middle of life's quiet rhythms, he saw something else.

Karma wasn't moral retribution. It was intelligent consequence. A ripple effect set in motion by every action—shaped by intent, circumstance, and energy.

Like physics, but alive. Like mathematics, but moral. Like causality, but conscious.

"Karma is not punishment," he wrote in his notebook later that evening. "It is the echo of intelligence acting through space and time."

Every thought, every word, every choice—each one generated momentum. Some immediate. Some delayed. But all intelligent in their unfolding.

And the beauty was this: We weren't slaves to karma. We were participants.

Emotion: Energy in Direction

As Keshav sat later in meditation, he turned his attention inward—toward the one thing science had always struggled to decode: emotion.

In labs, emotions were reduced to chemical signals, neurotransmitters, and hormone flows. In therapy, they were classified, diagnosed, managed. In corporate circles, they were suppressed, seen as distractions from productivity.

But here, in this stillness, Keshav began to observe something new.

Emotions were not irrational noise.

They were currents. Energy flows aligned with perception, survival, experience, and memory.

Fear wasn't weakness. It was intelligence detecting danger. Grief wasn't dysfunction. It was intelligence acknowledging loss and recalibrating identity. Joy wasn't frivolous. It was intelligence signaling alignment with one's deeper nature.

"Emotion is not the opposite of logic," Keshav realized. "Emotion is logic felt by the body before the mind understands it."

To deny emotion was to sever half of one's intelligence.

To fear emotion was to fear one's own compass.

Free Will: Conscious Redirection of Intelligence

The final piece fell into place one rainy morning as Keshav watched monks gather water in silence.

He had spent months exploring whether free will existed—or whether human life was just a programmed cascade of genetics, environment, and past karma.

And now he saw the truth:

Yes, we are shaped by karma. Yes, we are influenced by emotion. But within that flow, we can choose.

We are not free *from* consequences. But we are free *within* them.

Free will, he understood now, was not about breaking all laws of nature. It was about becoming conscious of the motion—and choosing to steer.

Just like a surfer doesn't create the wave... but learns to ride it.

Free will is the act of intelligent awareness redirecting its path.

It is intelligence realizing itself—and choosing to shape the next ripple.

The Integration: All Is Intelligence in Motion

It all connected now—like the final equation in a grand unified theory of life.

- Karma was the trail of intelligent choices.
- Emotion was the fuel and feedback of intelligent experience.
- Free will was the freedom within intelligence to evolve direction.

These were not mystical illusions. These were layers of the same living force—manifested through form, circumstance, perception, and choice.

"Intelligence is not static," Keshav wrote. "It is in motion—constantly becoming, constantly adjusting. We are that motion. We are that becoming."

In that realization, science was not rejected. Faith was not blindly accepted.

Instead, both were reconciled—within the living dance of intelligence.

Not human intelligence. Not artificial intelligence. But existential intelligence. Conscious energy in motion.

And That Changed Everything

Keshav was no longer searching for answers. He was living the answers.

No longer debating God. But recognizing divinity in motion.

No longer resisting karma. But seeing it as the canvas of conscious life.

No longer suppressing emotion. But honoring it as the voice of his own evolution.

And no longer fearing free will. But embracing it as the power to participate in the unfolding of creation.

He was no longer just observing intelligence. He was becoming it.

Aware. Balanced. Alive.

God Is No Longer a Myth—But a Living Reality

Keshav once believed that God was a concept. A cultural artifact. Something people needed when they lacked understanding—primitive comfort for a complex world.

In his younger years, he had debated religious scholars, dismantled theological assumptions with sharp logic, and dismissed sacred texts as poetic metaphors.

He wasn't cruel about it. Just... confident.

Confident that the universe could be explained by equations. Confident that reality didn't need a divine author—just better instruments.

But now, after months of deep introspection, observation, and humble surrender to experiences that refused to fit into formulas, he was no longer so sure.

In fact, he was sure of something else entirely:

God is not a myth. God is the most real thing he had ever encountered.

From Symbol to Substance

Keshav stood once again before a temple, but this time he didn't look at the deity as a statue. He looked at it as a symbol—a portal to something invisible, but present.

The chants, the offerings, the rituals—he once thought them hollow. Now, he saw them as human attempts to attune oneself to a greater intelligence.

And this realization wasn't born of belief. It was born of recognition.

The same force that created galaxies and governed quantum fields...The same intelligence that pulsed in his cells, his breath, and his thoughts... The same awareness that allowed him to observe reality...

That force—when directed with intention—was what people had always called God.

Not a man in the sky. Not a personality. But conscious energy with will. Intelligence that chooses.

That was divinity.

And he had been interacting with it his whole life without knowing its name.

God Is Not Outside. God Is Within the Pattern.

He no longer saw God as external. Not as an overseer watching from beyond the cosmos.

God was in the pattern itself. In the motion. In the pause. In the silence. In the becoming.

In the way water curves around a stone. In the way a baby cries the first time it breathes. In the moment of insight between two seemingly disconnected thoughts.

God was not somewhere else. God was what made everything here. And still, somehow, remained more than it.

This didn't require temples. This didn't reject science.

It revealed science to be the study of God's structure. And faith to be the surrender to its presence.

No Blind Belief. Only Living Understanding.

Keshav was clear: he had not turned religious. He had not embraced blind devotion or dogma.

He still questioned everything. He still held logic in high regard.

But what changed was this: He now understood the limitations of logic. He saw that some truths were not to be explained, but to be experienced.

"God," he wrote in his journal, "is not something to prove. God is something you begin to notice once your arrogance leaves the room."

He did not bow out of fear. He did not chant for reward.

He simply sat still— and felt the intelligence that was always watching, guiding, waiting.

And that feeling wasn't belief. It was intimacy.

When Intelligence Becomes Divine

What struck him most was this realization:

It is not intelligence that makes something divine. It is the awareness of intelligence with intent.

A tree grows—that's intelligence. A bird migrates—that's instinctive intelligence. A human reflects on suffering—and chooses compassion.

That's divine intelligence.

Because it is aware. It is free. It is choosing to align itself with something beyond its own survival.

That, he realized, was the missing piece all along. Divinity is not power. Divinity is intelligence made conscious—and directed with balance.

The Universe Was Never Empty. He Had Just Been Blind.

Keshav remembered his nights staring at stars through massive telescopes. Back then, he had thought he was peering into the emptiness of space.

But now, he knew: He had been peering into intelligent stillness. Into active silence. Into the canvas upon which the divine paints in atoms, light, time, and breath.

He hadn't been wrong in his science. Only incomplete.

He had studied the grammar of the universe. Now, he was beginning to understand its poetry.

This Was Not the End of His Search. It Was the Beginning of Real Vision.

God was no longer a question in his mind. God was a reality in his breath.

A presence in every decision. A tension between every action and its consequence.

God was not something he sought to meet. God was something he sought to mirror.

To live with intelligence that was:

- Aware,
- Balanced,
- Compassionate,
- Free,
- And aligned with truth.

Because if intelligence could create stars... it could also choose kindness.

And that choice—that is where God truly lives.

The Ultimate Realization

Keshav sat under the banyan tree in silence, the same silence that once used to feel empty—void of data, dialogue, or definitions.

But now, the silence was full. Not with noise or knowledge, but with presence.

It was the end of a long journey. Not the kind that measures miles, but the kind that erodes egos. The kind that strips a man of his titles, his theories, his carefully constructed worldviews—and returns him, naked and unarmed, to the mystery of being alive.

And in this vulnerable stillness, he finally understood.

No Longer a Seeker of Answers

All his life, Keshav had been taught that wisdom meant having the right answers. Answers were power. Answers were control. Answers filled research papers, solved equations, won debates, earned respect.

But now, as he looked back on the arc of his own transformation, he realized:

"The deepest truths don't come in the form of answers. They live inside questions—questions you carry like a mirror, not a weapon."

He no longer obsessed over solving God, consciousness, karma, or creation. He no longer needed to resolve every paradox or eliminate every uncertainty.

Because life itself was not a problem to be solved— It was a conscious phenomenon to be witnessed, participated in, and shaped—gently, wisely, intentionally.

The more he had tried to "solve" life, the more distant it had felt. The moment he began to live its questions—everything began to unfold, slowly, humbly, and beautifully.

Perception, Not Belief

For Keshav, the idea of God had always felt like a belief system—a binary switch: either you accepted it or you didn't. That's what religion had taught him. That's what atheism had countered.

But now, he saw clearly:

"God is not a belief. God is a way of seeing."

It was not about believing in a creator. It was about perceiving intelligence as sacred. Perceiving life itself as a miracle, not a mechanism.

You didn't have to pray to access it. You didn't have to reject logic to feel it.

You just had to become quiet enough, present enough, honest enough.

To remove the filters of conditioning, arrogance, fear, and distraction— And then, the divine revealed itself. Not in lightning. Not in booming voices. But in the quiet intelligence behind a breath, behind a thought, behind a choice.

Responsibility, Not Dependency

Another shift struck him deeply.

In the early days, the idea of God was deeply linked to dependency. People prayed for favors. They surrendered not out of awareness, but helplessness.

But now, Keshav saw that the presence of divine intelligence did not remove responsibility— It deepened it.

If intelligence was real, if it flowed through you, if it empowered your choices— then every act, every ripple, every intention mattered.

God was not watching to judge. God was flowing through you to evolve.

And the real surrender was not asking for miracles. It was asking,

"How do I align with the intelligent flow of life in a way that reduces suffering, spreads clarity, and honors the interconnectedness of all things?"

That was the new responsibility. And it was sacred. And it was yours.

Awakening, Not Arrival

There was no final enlightenment. No fireworks. No divine applause. No stamp of "now you know."

Keshav understood: awakening was not an event. It was an ongoing alignment. A continuous refinement of perception, presence, and purpose.

You don't wake up once. You wake up again and again—Each time you recognize the illusion, Each time you respond with awareness instead of reaction, Each time you choose stillness over control, Compassion over pride, Balance over ego.

And in those choices, God is lived. Not reached.

And Then—Peace

It didn't come like thunder. It came like rain. Slow. Quiet. Gentle. Filling every crack. Washing every corner.

Not the peace of escape, Not the peace of victory, But the peace of integration—

The peace of seeing that nothing was missing. That the universe was not broken. That he was not lost. That the questions he feared most were not enemies—they were invitations.

Peace, not as a passive silence. But as a living, breathing recognition:

"I am part of the intelligence. I am part of the becoming. And I am finally in rhythm with it."

The Final Realization

Keshav once believed intelligence was a tool. Now he saw—it was the very soul of existence. And when that intelligence becomes conscious of itself...it becomes what we've always called God.

Not in mythology. Not in debate. But in motion. In presence. In you.

He had not found all the answers. But he had found the place where they didn't matter anymore.

He had found peace. And perhaps, that was the ultimate truth intelligence ever wanted him to see.

10: The Decision – Breaking Free

"You don't just walk away from the world. You walk into a version of yourself the world can no longer contain."

The room was silent, but not empty.

Stacks of reports, half-sipped coffee, glowing monitors—everything around me still spoke the language of the life I once mastered. But for the first time, I wasn't listening.

Just days ago, I was in Varanasi, where a single realization had undone decades of belief. Not a mystical moment, not a divine miracle—just clarity. That God is not a figure. Not a force. But the motion of intelligence itself—when it enters form and acts with intent.

The Quiet Before the Storm

Now back in New York, standing inside my penthouse overlooking the skyline, I saw everything differently. The skyline looked like a simulation. My awards—tokens of a dream someone else planted. My reputation—a well-fed ghost.

I was not angry. Not lost. But undeniably done.

The scientist in me was not dying. He was evolving—shedding skin, cracking armor, burning away borrowed truths.

I didn't hate the world. But I could no longer pretend to be a part of its illusion.

What I was about to do would look like quitting. But it was the beginning of the **most radical experiment of all**: To live by awakened intelligence, not inherited algorithms.

And so, I reached for the last thing I'd write in this identity—not a research paper, not a patent, not another quarterly report. A letter.

To the world. To myself. To the system I no longer feared.

The Final Trigger

It wasn't the first betrayal. But it was the one that burned loud enough to silence everything else.

The email came at 6:43 AM—subject line: *"AI Integration Acceleration—Confidential."* I was looped in, almost by mistake. A junior executive had forwarded it for "technical review," not realizing what it contained.

It was a proposal—no, a decision. The board had already approved it. A new AI protocol was to be deployed globally through our satellite systems—not to monitor climate or improve connectivity, but to track human behavioral signatures. Thought patterns, micro-decisions, emotional variances—all in real time.

It would start with "smart cities," then be licensed to governments. "We are entering the age of predictive humanity," the note read. "As always, Keshav, we trust you'll ensure the science behind it is rock solid."

No one asked me. No one needed to. They knew the equations would work. That the code would run. That I'd done the impossible before—and would again.

But something snapped.

Not in anger. Not in fear. But in something deeper—alignment.

For months, maybe years, I had watched intelligence being commodified—used to manipulate desire, market addiction, model war, track dissent. But this was different. This was the weaponization of understanding itself.

And they were asking me to make it flawless.

I remembered a conversation I had years ago with an old professor in Madras. He had once told me:

"Be careful when your intelligence becomes a tool. Because one day, someone else will pick the target."

Now, the target was everyone.

I stared at the screen. My hands weren't trembling. My breath wasn't racing. But something inside me had stopped cooperating.

I walked into the next meeting and saw their faces—ambitious, brilliant, hungry. They spoke of scale, efficiency, vision. They quoted philosophers they didn't understand, ethics they didn't believe in, and used the word "humanity" like a marketing slogan.

I excused myself midway. Told them I needed coffee. I never returned.

Instead, I went home, opened my laptop, and pulled up the draft of my resignation letter. But instead of typing, I stared.

This wasn't about leaving a job. This was about leaving a version of myself that had served its time.

So I didn't write a resignation. I wrote a declaration.

Letter to the World

It began as a resignation. But halfway through, I realized—resignation is what people write when they still believe in the structure they're leaving. I didn't.

So instead, I wrote a letter. Not addressed to HR. Not even to my CEO. But to everyone who, like me, had once believed that intelligence alone was enough to build a meaningful life.

Subject: *This Is Not a Goodbye. It's a Wake-Up Call.*

> To whom it may concern— and it *should* concern every one of you,
>
> I am not resigning. Because you cannot resign from something that no longer exists. The person you hired—the scientist, the prodigy, the data evangelist—is gone.
>
> He didn't die. He evolved.
>
> For years, I helped build the most powerful systems known to man—networks that could learn, predict, manipulate, and scale. I was praised, promoted, and paraded as a symbol of progress.
>
> But here's the truth: we mistook capability for consciousness. We called it intelligence when it could out-calculate. We called it innovation when it could out-market. But what we never stopped to ask was—*intelligent for what*?

I have watched our brightest minds optimize addiction, forecast obedience, and simulate morality. I have watched your algorithms reshape elections, rewrite emotions, and rewire entire generations to need what they never wanted.

And I was part of it.

Not out of malice. Out of belief.

I believed that logic would lead to truth. That efficiency would lead to freedom. But it doesn't. Not without intention. Not without alignment. And certainly not without awareness.

So this is not an apology. This is a confession of awakening.

I am leaving because staying means silence. And silence now is complicity.

I no longer wish to be an enabler of unconscious progress. I want to become a disruptor of blind momentum.

I won't be joining a competitor. I won't be founding a startup. And no, I'm not retreating into the mountains.

I am becoming a monk. Not one who chants in caves, but one who walks through cities. Asking questions. Disturbing comfort. Breaking illusions.

Because the future doesn't need more products. It needs more perception.

And the world doesn't need another system. It needs a mirror.

If this letter makes you uncomfortable—good. It means your intelligence is still alive.

With awareness,

Keshav Raghunathan

—A scientist no more. A seeker, forever.

He didn't send it as a PDF. He published it on his blog, then sent the link to everyone on the global listserv. Then he closed his laptop and packed his bags—not for exile, but for entry. Not to disappear, but to reappear where it mattered most.

Not an Escape, but a Stand

People assumed I was running away.

Some whispered it—"burnout," "midlife crisis," "existential breakdown." Others offered polite sympathy, like I was grieving a personal loss. A few even tried to stage interventions, as if I were on the verge of madness.

But none of them understood.

Because leaving *this* world doesn't mean you've lost your mind. Sometimes it means you've finally found it.

I wasn't retreating into a forest. I wasn't shaving my head or wrapping myself in saffron. I wasn't rejecting modern life—I was rejecting unconscious life.

"This is not renunciation," I told them. *"This is rebellion—against automation of the soul, against the outsourcing of awareness."*

When you're born into a system, programmed by its definitions of success, it doesn't just shape your future—it hijacks your identity. You start mistaking performance for purpose, achievements for alignment, and busyness for meaning.

I had lived that life. And excelled at it.

But what good is success if it's built on a sleepwalking mind? What good is intelligence if it's leased out to agendas you never questioned?

This wasn't me walking away from responsibility. This was me finally taking it.

The real escape would have been staying—collecting bonuses, publishing white papers, building machines smarter than their makers, and pretending it was all for the "greater good."

No.

That would have been escape. That would have been cowardice dressed in prestige.

I chose confrontation. But not with weapons. With awareness. Not with anger. With clarity.

To confront unconsciousness wherever I saw it—boardrooms, classrooms, temples, or protests. To question not just the systems we build—but the self that obeys them.

Because you can only awaken the world by first refusing to be its dreamer.

Burning the Name Tag

It was just a piece of plastic. Black letters on a silver background: Dr. Keshav Raghunathan Director – Strategic AI Initiatives

I had worn it for years—at conferences, global summits, press interviews. It had opened doors, turned heads, and earned respect before I even spoke a word.

But now, in my hands, it felt absurd. A lie engraved in Helvetica. A name tag not of who I was, but of what I had become to the world.

The titles, the degrees, the designations—they weren't badges of honor anymore. They were masks. Masks that kept me aligned to expectations, not truth.

I stepped onto the balcony, the winter air wrapping around me like a silent witness. Below, the city kept glowing—its endless hunger shining through skyscraper glass.

I held the name tag in one hand, a lighter in the other.

At first, I hesitated. Not out of doubt, but out of reverence. Not for the institution. Not for the achievements. But for the version of me who had worked so hard, thinking it was the way.

"You served your purpose," I whispered. *"But I can't carry you any further."*

I flicked the lighter. The flame caught. The plastic warped, the edges curled, and the name—*my name*—melted into nothing.

I watched in silence. Not with sadness. But with release.

That tag had once represented *arrival.* Now it marked *departure.*

I didn't stop there.

I cleared my office. Deleted my personal bio from the company server. Wiped my cloud storage. Archived everything under a single folder: "Former Self."

Then I wrote one last journal entry, not as a scientist but as a witness:

"Today, I erased the name the world knew, so I could meet the one it never did. Not the one in articles or awards. But the one beneath all definitions—the one who observes, questions, and walks without a label."

From that moment on, I was no longer Dr. Raghunathan. Not a designation. Not a product of pedigree.

Just Keshav—a ripple in awareness, moving with intent.

The Monk Without Robes

I didn't shave my head. I didn't wear ochre robes. I didn't carry a begging bowl or chant Sanskrit mantras on Himalayan peaks.

There was no ritual, no initiation, no spiritual master welcoming me into an ashram. Because my monkhood wasn't a change of clothes. It was a change of conscious alignment.

People looked at me strangely in the days that followed. I wasn't dressed like a guru. I wasn't speaking in riddles. I still wore jeans and black T-shirts. Still took public transport.

Still walked through airports with a duffel bag and a single book in my hand.

But inside, everything had shifted.

Where I once analyzed, now I observed. Where I once reacted, now I inquired. Where I once sought control, now I sought resonance.

I was no longer trying to dominate the system. I was now trying to decode it—layer by layer, belief by belief, ego by ego.

I didn't want to teach people how to live. I wanted to help them see how they were living unconsciously.

I wasn't against science. I was against blind submission to it. I wasn't against faith. I was against faith without inquiry. I wasn't against the world. I was against the hypnosis of its momentum.

"A true monk," I wrote in my notes, "is not one who renounces the world— but one who sees it clearly, walks through it awake, and dares to speak what others are afraid to see."

That was now my path.

I didn't need robes to feel sacred. I didn't need rituals to feel connected. Every breath was a chant. Every question was a prayer. Every moment of stillness—my temple.

People expected silence from monks. But I had chosen a different kind of silence—not the absence of speech, but the absence of noise within.

I had become a monk of the mind, armed not with scripture, but with clarity.

And the first illusion I was ready to challenge was the one I had worn for decades: That knowledge is wisdom. That achievement is understanding. That intelligence is truth.

Calling the Unknown

There comes a moment when your next step has no roadmap. No mentor, no model, no precedent. Only a pull—a silent, undeniable force that says: "Now. Go."

For me, that moment arrived in the stillness after the fire.

The name tag was gone. The resignation letter had become a declaration. The illusion had collapsed. All that remained was truth—unpolished, uncomfortable, alive.

I opened my laptop, not to apply for another job, not to update my résumé, but to buy a one-way ticket. Destination: India. Arrival: Unknown.

No ashram awaited me. No family home felt like home anymore. No plan. No itinerary. Just the intent to walk forward without the burden of a lie.

I chose the cheapest red-eye flight. Middle seat. No upgrades. No comfort. Because this wasn't a homecoming. It was a re-entry into reality, stripped of privileges and stories.

I packed a small bag—two pairs of clothes, a journal, a pen, and a book I hadn't opened since college: *The Bhagavad Gita.* Not for answers, but for resonance.

I looked around my apartment one last time—cold, curated, successful. It had been my fortress. But now it felt like a museum for a man who no longer existed.

I left everything behind—furniture, gadgets, nameplates, trophies, even my meticulously organized bookshelf. Because even knowledge can become baggage when it's disconnected from the soul.

When the cab arrived, the driver asked casually, "Heading somewhere special?"

I smiled.

"No. Just finally heading somewhere true."

I didn't know where I would land once I reached India. Not physically. Not emotionally. Not intellectually.

But I knew one thing with unshakable certainty:

I would no longer live a borrowed life. I would no longer speak words I didn't believe. I would no longer measure myself by systems designed to keep me asleep.

For the first time, I wasn't chasing a destination. I was answering a call.

Not from a place. But from a part of me that had waited patiently behind all the noise.

The unknown didn't feel like a threat anymore. It felt like home.

The Message He Leaves Behind

News of my exit spread faster than I expected. Within hours of publishing the letter, my inbox was flooded—subject lines swinging between concern and confusion.

"Are you okay?" "What happened?" "Is this a hack?" "Are you joining a cult?" "You're too smart for this, Keshav."

Some colleagues tried to be diplomatic. Others didn't bother.

I received one message from a former mentor that read:

"Genius and madness have always shared a thin line. I hope you're not crossing it for good."

I replied:

"If madness means no longer being a slave to false success, then yes. Finally, I'm free."

The responses ranged from disbelief to pity.

One junior researcher wrote:

"Sir, I joined this field because of you. Your TED talk changed my life. Why are you abandoning us?"

I didn't respond immediately. I stared at his message for a long time, heart heavy—not with regret, but with recognition. I saw my younger self in him, eyes full of dreams, unaware that he was climbing a staircase that looped back to the same floor.

So I wrote back:

"I'm not abandoning the mission. I'm abandoning the illusion. Real science isn't about machines that win. It's about minds that wake up. And if I can help just one person truly question what they're chasing, then I'm still doing the work. Just not in a lab coat anymore."

Some appreciated that. Others rolled their eyes. And a few, silently, began to wonder.

The head of HR called to "check in." I let it ring. The CEO left a voicemail offering a "sabbatical and reflection leave." I deleted it.

Because this wasn't a burnout. It wasn't a phase. It wasn't a career pivot.

It was a conscious exit from the game of pretending.

No security escort. No farewell party. No framed memento. Just a man walking out with nothing but clarity—and leaving behind a silence loud enough to echo in the minds of those still trapped inside.

One colleague finally called and asked directly, "So what now, monk?"

I smiled.

"Now... I start disturbing minds instead of machines."

The Road Begins Where the Identity Ends

There was no music when I walked out. No dramatic goodbye. No sunset to ride into. Just the soft hum of an elevator descending from the 47th floor—carrying someone who was once a scientist, an innovator, a celebrated mind.

Now, I was just... me.

And strangely, for the first time, that was enough.

I stepped onto the street as the city buzzed around me—horns blaring, screens flashing, people glued to devices they mistook for direction. No one noticed a man leaving everything behind. But that was the point.

Transformation doesn't always look like a lightning strike. Sometimes, it's just the quiet defiance of walking away from what no longer feels true.

I wasn't leaving the world. I was entering it, finally awake.

Not to escape. But to listen differently. To ask questions that would unsettle boardrooms and meditation halls alike. To step into chaos—not to fix it, but to understand why it exists in the first place.

There was no map. But I didn't need one.

The identity I once built—layer by layer, degree by degree—had served its purpose. It had taken me to the edge of meaning. But it could not cross over.

Only awareness could.

As I walked through the airport later that night—no VIP lounge, no corporate handler, no return ticket—I carried nothing but intention.

People passed by me with purpose in their eyes. But I wasn't in a rush anymore. Because this time, I wasn't going somewhere to become someone.

I had already become no one—and in that, I had found everything.

I thought about all the monks I had read about as a child. Their solitude. Their robes. Their quiet.

And I realized something.

Monks are not the ones who leave the world. They are the ones who enter it fully awake— with nothing to prove, nothing to protect, and nothing to hide behind.

And so, with no forwarding address, no job title, and no plan, I took the first step onto a road not built on ambition—but on awareness.

Not a pilgrimage. Not an escape. Just the beginning of a long, necessary conversation with the world.

11: Becoming "The Monk"

"To become a monk is not to disappear. It is to become visible in a way the world forgot how to see."

Transformation is never marked by applause. It happens in silence—just after you've left everything behind, but before the world knows who you've become.

Keshav had crossed that threshold. He was no longer a man defined by achievements or rebellion. What remained was not an identity, but an intent—clear, stripped, and steady.

He had not found a new label. He had found a new way of being.

And that is where this new chapter begins—not with answers, but with the awareness that questioning itself is sacred.

Shedding the Last Layer - The Birth of a Name

The world loves titles. They help people understand where to place you—what shelf to put you on, what kind of respect to give, what kind of distance to maintain.

For most of my life, I had one: Dr. Keshav Raghunathan. Scientist. Innovator. Speaker. Visionary. Every word after my name was an instruction—this is someone who "knows."

But now, I knew better.

Knowledge is not who you are. It is what you carry—until you realize most of it was handed to you by someone else.

As I prepared to step into a life without a formal role, I felt the weight of even my name. It wasn't the sound of it that bothered me—it was what it had come to represent. A brand. A legacy. A product polished by institutions and positioned by achievements.

I considered adopting a spiritual name. Something Sanskrit. Something sacred. I even tried a few in my journal: *Advayananda*, *Tatvadarshi*, *Anubhavananda*. They all sounded correct—but none of them felt true.

Because I wasn't trying to become a symbol. I was trying to become transparent.

And so, I kept the name the world already knew—Keshav Raghunathan. But this time, I added no titles. No degrees. No achievements.

Just one phrase beneath it:

Seeker of Intelligence

Not a guru. Not a teacher. Not a dropout. Just a ripple in the field of awareness—trying to align, trying to understand.

"I am not here to lead or to follow. I am here to dissolve what I was not, and see what remains when there's no one left to impress."

That was my vow.

In a world obsessed with labels and lineage, I chose intent over identity. Not because I was beyond the world—but because I finally wanted to walk through it *without filters.*

And in doing so, I didn't lose who I was. I discovered it—for the very first time.

Not a Teacher, but a Questioner

Everywhere I went, people asked the same thing:

"So now that you've awakened, what will you teach?"

They wanted content. Methods. Models. A framework for truth. Some even wanted a syllabus.

But I had no teachings to offer. No "five-step enlightenment plan." No spiritual shortcut dressed in philosophy.

Instead, I had one sentence echoing in my mind with absolute certainty:

"I am not here to teach. I am here to make people question."

Because the moment you start teaching, the world starts believing you have answers. And the minute people believe you have answers, they stop looking for their own.

I wasn't here to transfer wisdom. I was here to ignite inquiry—the kind that burns through comfort, challenges borrowed beliefs, and leaves people staring into the raw mirror of their own thoughts.

Teaching, I realized, was often just sophisticated influence. It replaced one set of instructions with another. And influence, no matter how noble, is still control—a quieter domination of someone else's mind.

That's not what I wanted.

I didn't want followers. I didn't want students. I didn't want believers.

I wanted people to catch themselves mid-sentence, mid-thought, mid-assumption—and ask: *"Wait... why do I think this way?"*

That is the only worthwhile beginning.

I stopped preparing talks. Stopped drafting lectures. Instead, I began carrying questions like tools:

- "What if the things you believe are the reason you suffer?"
- "Whose voice speaks in your head when you call it 'truth'?"
- "Is your ambition yours—or was it inherited, disguised as desire?"

I didn't want to impress. I wanted to disturb gently.

If I could mirror reality so clearly that someone saw their distortion for the first time, that was enough. That was the real transmission.

"I am not a light," I wrote in my journal. *"I am the crack in the wall—through which your own light escapes."*

This path is not glamorous. There are no applause breaks for asking difficult questions. But truth is never loud. It is usually whispered—inside the pause between your conditioned thought and your first honest doubt.

Intelligence Becomes the Compass

I didn't throw my science away. I simply took it off the pedestal.

The microscope, the neural networks, the quantum formulations—they weren't the problem. The problem was the *worship*. The arrogance that came with knowing how the engine runs, while forgetting *why* we built it in the first place.

For years, intelligence was treated like data mastery—an ability to compute faster, analyze deeper, or debate louder. But that version of intelligence felt... thin. Hollow. Loud on the surface but fragile at its core.

Now, as I stood with nothing to prove and no institution to defend, I began seeing intelligence not as a performance metric, but as an alignment mechanism.

A ripple in still water doesn't shout. It simply responds—naturally, proportionally, in tune with the disturbance.

That, I realized, is what true intelligence does. It doesn't overpower reality—it aligns with it. It senses the field, adjusts its frequency, and responds *intuitively* to what is.

One evening, sitting by a railway platform in India, watching people move with unspoken patterns of survival and chaos, I opened my journal and wrote:

"Science without intelligence is blind. Faith without intelligence is superstition."

Both systems—science and faith—had failed in the same way: They had become obedient.

Science had surrendered to models. Faith had surrendered to dogma. Both had stopped listening.

Intelligence was the missing link.

It is not belief. It is not logic. It is the capacity to resonate with what is truly happening, moment to moment, beyond one's conditioning.

It doesn't belong to any system. It doesn't wear a lab coat or a robe. It doesn't speak in formulas or verses. It *observes*. It *questions*. And when necessary—it *interrupts*.

I began to see how every human contradiction—between science and spirit, logic and love, control and surrender—was not a real paradox. It was just a gap in intelligence.

Fill that gap, and everything begins to make sense—without force.

I didn't need to reject my past. I just needed a better compass. Not one that pointed north. But one that pointed inward, toward alignment.

Preparing for the Path

You don't need much to walk a real path. In fact, the less you carry, the clearer the road becomes.

In the first few days of my return to India, I sat alone in a quiet room at a friend's guesthouse in Rishikesh. There was no Wi-Fi. No agenda. No one expecting me.

And that's when I began to take inventory—not of what I had, but of what I no longer needed.

What I released:

- **Applause:** The constant need to be validated, quoted, praised, or retweeted. Truth doesn't echo. It *resonates silently*.
- **Titles:** Doctor. Director. Scientist. Visionary. Labels that once opened doors but now only

blocked the mirror.

- **Debates:** I no longer wanted to win arguments. I wanted to understand *why* we argue.
- **Defenses:** My past choices, my knowledge, my philosophies—I didn't need to guard them anymore. If something was true, it would stand on its own.

In their place, I gathered a few quiet tools—subtle, powerful, and enough for a lifetime:

- **Observation:** To see things *as they are*, not as I'm told they are.
- **Silence:** Not just the absence of noise, but the space where truth sneaks in unannounced.
- **Inquiry:** Not to reach conclusions, but to dissolve illusions. A question asked deeply is often worth more than a thousand answers recycled.

Most people prepare for the future by adding. I was now preparing by subtracting.

No plans. No lecture circuit. No marketing. Just clarity.

I remember writing this one line across an entire page in my journal:

"I know less than ever. But I am clearer than I've ever been."

It was paradoxical. And it was perfect.

Because the kind of knowing I had now wasn't about information. It was about attunement—to reality, to people, to myself.

In the world I came from, people prepared for success. In the world I was entering, I was preparing for truth—and that needs no rehearsal.

The First Step as "The Monk"

There was no red carpet waiting. No crowd to welcome me back. No reporters asking, "Why did the AI visionary vanish?"

Good.

Because I didn't return for spectacle. I returned for stillness.

My flight had landed in Dehradun early that morning. I had no hotel booked. No event to attend. No destination marked on a map.

I boarded a rickety local bus heading toward the unknown—just a name I liked on the signboard: *Devprayag*. It didn't matter where I ended up. For the first time in my life, I was not chasing anything.

As we neared the edge of the small town, the bus dropped me near a railway crossing. I stepped out with a single backpack and a worn-out pair of sneakers—more monk than CEO, even if no one around me knew.

And I walked.

Past temples layered in incense and bells. Past chai stalls where old men debated politics in the same tones I once used to debate machine learning models. Past children laughing without purpose. Past street dogs fighting without hesitation. Past women bargaining for vegetables with intensity that made Wall Street look like kindergarten.

I walked through it all, not as a visitor, not as a savior—but as a listener.

Listening not just to the town's rhythm, but to my own reactions. The subtle judgments that still arose. The pride that still clung to humility. The need to make sense of things too quickly.

I caught myself wanting to speak. To explain who I was. To offer insights.

But then I remembered:

"You are not here to teach. You are here to unlearn."

So I said nothing. I simply observed—like the scientist I once was, but now without the ego of knowing.

Later that evening, sitting on the stone steps near the confluence of rivers, I scribbled one clear sentence in my journal:

"I didn't come back to India to find a mission. I came to dissolve the illusion that I ever needed one."

There was no epiphany. Just a soft, grounded beginning.

Not as a monk with answers. But as a seeker walking barefoot through his own layers of illusion.

And that—finally—was enough.

Part IV: The First Step Toward Truth

(The beginning of his world tour and first seminar)

12: The First Invitation

A University Debate

"When the world expects a scientist and a monk shows up, something deeper than curiosity is stirred—something primal: the need to rethink everything."

The monk had stepped into the world. Not to retreat from it, but to disturb its assumptions.

Keshav was no longer a silent seeker wandering through railway stations and riverbanks. His presence had begun to ripple—quietly, invisibly—until it reached the corridors of one of India's most prestigious universities. An invitation arrived, addressed to the renowned AI visionary, unaware that the man they hoped to welcome no longer existed.

They were expecting answers, theories, and dazzling predictions about the future of machines. What they got instead was a man who had unplugged from the very system he once mastered—ready not to teach, but to **make them question**.

This was the first moment the world would see him not as a defector, but as **a mirror**. And as he would soon discover, mirrors are rarely welcomed in places built on appearances.

The Unexpected Invitation

The email subject line read: "Keynote Invitation: AI & Consciousness – Annual Tech Symposium, IISN."

IISN. One of India's top science and technology institutes. A place that had once quoted my papers. Cited my patents. Shortlisted me for honorary awards I had never bothered to accept.

I stared at the message for a long time.

They were inviting Dr. Keshav Raghunathan, the renowned scientist. The mind behind emotion-mapping algorithms and sentient machine protocols. The man who had vanished from the global stage without a word.

They had no idea who they were writing to now.

The invitation was polite, formal, and enthusiastic:

"We would be honored to host you as the keynote speaker at our annual symposium. Your pioneering work in artificial intelligence and neural interfacing has shaped a generation of researchers. This year's theme—'AI and Consciousness'—would be incomplete without your insights."

I almost declined it immediately.

Not because I feared the stage. But because I feared being misunderstood. Again. They were expecting brilliance wrapped in certainty. What I now carried was awareness wrapped in humility.

I closed the laptop, walked out into the noise of the town, and let the question breathe:

Should I go back—not to belong, but to confront? Not to debate, but to mirror?

The next morning, in a simple moment of stillness, I had my answer.

Yes. Not to lecture, but to interrupt.

I replied to the email with a short note:

"Thank you for the invitation. I accept. But please note: I no longer carry slides or deliver predictions. My presence is not to explain AI, but to explore what remains beyond it."

No one replied for a day.

Then came a hesitant confirmation. A changed title: "Keshav Raghunathan – Special Guest: A Journey Through Intelligence."

They still didn't know what was coming.

Neither did I. But I knew one thing:

This was not a return to science. This was my first step into the heart of its illusions.

The Arrival – Confusion in the Hallway

The auto-rickshaw dropped me just outside the institute's main gate. No VIP escort. No banner with my name. No one recognized me.

Perfect.

I walked past the security desk unnoticed, carrying nothing but a cloth shoulder bag and a notebook worn soft by travel and thought. My clothes were plain—white cotton kurta, loose pants, old sandals. I had shaved, not for appearance, but for shedding.

Inside, the atmosphere was electric. Projectors humming. Volunteers rushing. Students clustering around QR-coded posters about machine ethics, artificial sentience, and the boundaries of consciousness.

I could feel the expectation in the air. They were waiting for someone else.

Near the main seminar hall, two junior organizers stood with printouts in their hands. I walked up to them.

"Excuse me," I said. "Can you point me to the speaker lounge?"

They glanced up, distracted. One of them looked at me again and blinked.

"Um... Are you... here for... Dr. Keshav Raghunathan's session?"

I smiled gently.

"Yes."

They paused. You could almost hear the thought loop short-circuit: *Wait. This can't be him. No badge. No formal shoes. No laptop. No assistant.*

One of them stepped aside and tapped on her phone furiously. The other whispered to a nearby coordinator:

"Is that really him?"

Their reaction didn't bother me. It revealed exactly why I was here.

They weren't expecting a presence. They were expecting a presentation.

They had invited a mind. They got a mirror.

As I stepped into the green room, I saw a table laid out with bottled water, technical briefings, and an HDMI connector waiting for my laptop.

There was no laptop.

Instead, I sat in the corner, closed my eyes, and let the moment settle.

I wasn't here to dazzle them. I was here to dissolve something invisible— the illusion that appearance equals authority, that intellect equals identity, that knowledge alone can awaken.

Outside the door, the audience was buzzing. The lights were being adjusted. The microphone was tested.

I could hear a student say:

"He better have something big. We've been hyping this for weeks."

I exhaled softly and stood up.

Let the silence begin.

The Opening Speech – Breaking the Frame

The auditorium lights dimmed. Applause began out of habit, not recognition. Most in the audience had only seen the man I used to be—on screens, in papers, quoted in journals. Now they saw someone else.

A murmur rippled through the rows as I stepped onto the stage—no PowerPoint, no laser pointer, no branded suit. Just presence.

I looked out—not at the crowd, but into it. Noticing not just faces, but postures. Anticipation. Skepticism. Polite doubt.

The kind of silence that precedes either a revelation—or a rejection.

I adjusted the mic, took a breath, and began:

"Science tells us how. Religion tells us why. But neither tells us who we are."

The words hung in the air. No opening pleasantries. No technical background. No credentials.

Just a crack.

A deliberate fracture in the frame they had brought with them—the frame that divides disciplines, divides minds, divides truth.

I stepped back from the podium and continued, slowly:

"You invited me here as a scientist. Some expected formulas. Others feared philosophy. But I am here for something far more disruptive. I'm here to invite you—not to learn—but to see."

A few brows furrowed. Some pens stopped moving. Good.

"We live in an age where information is confused with intelligence. Where models explain everything—except the one asking the question. We build machines that simulate emotion, but still cannot explain our own grief. We debate the ethics of AI while ignoring the illusions of the 'I' that debates."

I paused. Not for effect, but for breath. To let the resistance rise.

Then I offered the key:

"This is not a rejection of science. And it is not a surrender to belief. This is integration—of science, spirit, and self. Because the future will not be built by better machines— It will be saved by clearer minds."

Silence again.

Not because they agreed. But because their frames were beginning to crack.

"I'm not here to give answers. I'm here to question the lens you use to absorb them. Because if your lens is fractured, even truth looks distorted."

And then, stepping slightly forward into the light:

"I am not standing here as a scientist, nor as a sage. I stand here as a human—who chose to become still long enough to ask: What if everything we think we know... is just a beautiful theory protecting us from inconvenient clarity?"

A pause. A ripple.

And with that, the first layer of the debate was no longer academic—it had become existential.

The Student's Challenge

The room was still—too still.

Some eyes sparkled with curiosity. Others squinted with quiet resistance. A few scanned their programs again, perhaps wondering if they had walked into the wrong session.

And then it came.

A rustle. A chair scraping the floor. A voice—young, sharp, and barely masking irritation.

"Are you saying science is... incomplete?"

All heads turned toward the third row from the front.

A student—early twenties, thick glasses, hoodie with a physics quote. The kind of mind that had spent years sharpening itself on equations, empirical logic, and a fierce pride in rationality.

I looked at him—not as an opponent, but as a younger version of myself.

"Because that's what it sounds like," he continued, gaining confidence. "That you're implying science is blind or broken. Isn't that dangerous? To plant doubt in minds that are finally learning to think clearly?"

Murmurs followed.

Some nodded. Others glanced at faculty members as if to gauge their reaction. A few smiled—not with agreement, but with anticipation. The room had come alive.

And in that moment, I could feel the tectonic tension beneath the polished floorboards:

- The old comfort of *certainty* brushing up against the raw edge of *awareness.*
- A system trained to answer, now being asked to *pause.*
- Logic, feeling threatened—not by ignorance, but by depth.

I walked away from the podium, closer to the front row, and looked at the student directly.

Not with challenge. But with respect.

Because this was the moment that mattered—not when people agree, but when they care enough to defend what they believe.

I smiled gently and said,

"That's a fair question. And it deserves an honest answer."

The room fell into a deeper silence. Not passive this time, but charged.

They didn't know it yet, but this wasn't going to be a debate. It was going to be a mirror.

And the first reflection was about to begin.

Keshav's Response – Clarity Over Comfort

I let the question hang a little longer— not to win time, but to give space.

You see, when people defend science, they're not just defending a method. They're defending identity. The idea that knowledge makes us powerful. That proof protects us from chaos. I understood that deeply—because I had once lived inside that armor too.

So I didn't answer with a counter-argument. I answered with a pause—and then, with this:

"Science can measure reality... but it cannot experience it."

The room didn't know how to react. Some blinked. A few leaned in. Most waited, unsure if this was the beginning of clarity or the end of credibility.

I continued.

"Faith, on the other hand... can give purpose—but it cannot explain intelligence."

Now the tension grew. Two worlds had just been touched—science and spirituality—and I had placed a question mark over both.

So I clarified:

"I'm not here to reject either. I'm here to unite what has been divided."

I walked slowly back toward the podium—not to elevate myself, but to let the idea rise.

"Science, when pure, is observation in motion. Faith, when pure, is surrender in stillness. But both have become corrupted—one by ego, the other by fear. One seeks control, the other clings to certainty. What is missing... is intelligence."

Another pause.

"Not artificial intelligence. Not emotional intelligence. But the intelligence that aligns—with reality, with awareness, with life itself. It is neither mechanical nor mystical. It does not belong to any lab or temple. It is what remains when the noise of method and myth fall silent."

Now the murmurs had quieted. Even the student who had asked the question was still.

So I looked at him again and said, not as a rebuttal, but as an invitation:

"Your question isn't wrong. It's just incomplete—because it assumes science is either everything, or nothing. But what if it's just a tool, not a truth?"

I stepped aside.

"And what if the only thing that can truly hold science and faith together—without breaking either—is the one thing we forgot to nurture: Intelligence that listens more than it concludes."

Reaction – Shock, Curiosity, Controversy

He didn't raise his voice. He didn't shout or condemn or debate. And yet—by the time Keshav stepped down from the stage, the entire auditorium felt like it had been cracked open.

Not by volume. But by vibration.

Some students rose to their feet in applause, not quite knowing *why* they were clapping—only that something inside them had shifted.

Others stayed seated, blinking in silence. One professor whispered to another,

"That wasn't science. That was... something else."

The Director of the institute, who had prepared a glowing closing remark, folded his note halfway through and chose not to speak at all.

Because how do you summarize something that refuses to be summarized?

By the next morning, the news had exploded:

The Scientist Who Became a Monk Stuns Tech Symposium with Philosophical Bombshell

Trending on social media:

#FromLabToLiberation

#WhoAreWeReally

#KeshavUnplugged

Clips of his speech were circulating on reels and academic WhatsApp groups alike. One 20-second clip went viral:

"Science can measure reality. But it cannot experience it."

Young minds, tired of binaries, found a new kind of oxygen. PhD scholars quoted him. Spiritual channels claimed him. Mainstream media tried to label him—and failed.

"Mystic in disguise?" "Anti-science alarmist?" "Spiritual reformer with a tech background?" "Post-rationalist monk?"

But he fit none of the frames. And that was exactly the point.

Back at the university, the organizing team was in disarray. The session that was meant to crown their academic prestige had turned into an unplanned awakening.

Some faculty members demanded an apology for "blurring boundaries between science and spirituality." Others defended the event, calling it "the most important conversation we've hosted in a decade."

Students began forming informal discussion circles, quoting his lines, debating meaning—not equations, but *existence*.

The symposium's closing ceremony, which usually included awards for technical papers, was overshadowed by whispers of "the monk," and what he had left behind—not facts, but friction.

What started as an honorary keynote became a catalyst for a movement.

And Keshav? He was already gone.

His Quiet Exit

No press photographs. No bouquet. No plaque or certificate.

Keshav stepped out of the auditorium the same way he had entered—unannounced, unnoticed by most.

The campus sky had turned to twilight. A soft wind stirred the fallen seminar posters scattered near the walkway. Behind him, the auditorium still buzzed with aftershock—the echoes of ideas too sharp to ignore, too fluid to contain.

He didn't stay for the closing dinner. Didn't pose for selfies. Didn't check the trending hashtags.

Instead, he walked across the campus lawn barefoot, letting the earth cool his feet. The same earth beneath research labs, temples, slums, and mountain peaks.

He passed by a group of students still debating. They didn't notice him. He smiled—not because he needed their recognition, but because he had already given them something more dangerous than knowledge: a question without a boundary.

As he reached the main gate, a thought rose—not loud, not poetic, but true:

"The world doesn't need more knowledge. It needs cleaner minds to hold it."

He didn't mean purity in the moral sense. He meant clarity—minds no longer hijacked by titles, dogmas, or borrowed ambition. Minds willing to observe without preloading judgment. To feel without drowning. To think without clinging.

The guard at the gate nodded at him, unaware he was holding it open for a man who had just disturbed the foundations of certainty.

Keshav stepped out.

No fanfare. No closure.

Just silence.

But inside that silence... a ripple had been released. And the world, whether it liked it or not, had already begun to change.

Epilogue: The Journey Begins

A week after the university debate, his inbox flooded. Not with promotional emails or project proposals. But with invitations.

From Ivy League universities and obscure liberal arts colleges. From Buddhist monasteries and European philosophy forums. From think tanks, dissenters, seekers, skeptics, and even tech CEOs who couldn't quite explain why his words made them uncomfortable—and yet, seen.

Some invitations were wrapped in curiosity. Others, in controversy. But all were united by one invisible thread:

The monk had said something they couldn't unhear.

And so, the tour began.

But it wasn't a tour in the conventional sense. There were no book signings, no stage lights, no brand managers.

Just Keshav. A backpack. A worn journal. And a mind stripped of identity, yet full of clarity.

He didn't go to preach. He went to disrupt alignment with illusion.

At every stop, he sat—not as a guest of honor, but as a mirror. He asked questions no curriculum had room for:

- *What is time, if not a tension between perception and*

causality?

- *Can reality exist without the observer?*
- *Why does intelligence feel imprisoned by the systems it creates?*

Audiences expected spiritual truths or scientific marvels. Instead, they got unsettling silence followed by precise, piercing thought.

As the global ripples widened, one student asked him at a cafe in Prague:

"What exactly are you trying to do? Start a movement?"

Keshav chuckled, stirred his tea, and replied:

"Truth is not something you believe in. It is something you realize."

"I do not seek followers. I seek thinkers."

This was no spiritual revolution. No anti-science rebellion. It was something more dangerous to every comfort zone:

A monk with no robes. A scientist with no lab. A mirror with no bias. And a mission with no end.

He didn't ask people to believe. He invited them to see—clearly, deeply, without distortion.

And that's where the first book ends.

Not with closure. But with a doorway.

Up Next: *Time, Reality, and the Observer*

The first global dialogues begin. Worldviews collide. Concepts are challenged. And Keshav enters the next battleground: Not the world outside—but the illusions held within every mind that claims to know "what is real."

The monk is on the move.
And so is the truth he came to reflect.

Final Thought

Volume 1 is the Genesis of the Entire Series

This is not a story of one man. It is the awakening of a mind that refused to remain programmed.

Volume 1 is not just about Keshav's transformation— It is the birth of a revolution that begins within, not outside.

He was once a celebrated scientist—armed with data, driven by reason, and applauded by the world. But something cracked. Not because life failed him, but because truth demanded more than intelligence—it demanded alignment.

And so, he stepped away. Not to renounce the world. But to re-enter it—without illusion, without identity, without fear.

In this first book, we witnessed:

- The unraveling of a mind once addicted to certainty.
- The collapse of definitions—of God, purpose, and self.
- The rise of a new way of seeing: where science meets silence, and faith meets inquiry.

He didn't switch professions. He changed direction.

From mastery to surrender. From performance to presence. From intellect to intelligence.

This is not a spiritual fantasy. Nor a motivational arc. This is a manual of liberation—disguised as a memoir.

Volume 1 ends where most stories begin: At the edge of the world.

Because now...Keshav steps out not just to explore truth, but to debate it, challenge it, and dissolve it—country by country, thinker by thinker, belief by belief.

The stage is no longer a lecture hall. It is the entire planet. And the dialogues ahead will be far more fierce, diverse, and unfiltered.

What You're Holding Is the Genesis

- This book is not just a transformation story— it is the blueprint for the journey to come.
- Keshav awakens to the truth about intelligence, God, and free will— not as abstract ideals, but as inner realignments.
- His journey from scientist to monk is not an escape— it is an evolution of responsibility.
- And this is just the beginning— the real test starts now.

In Volume 2: *Time, Reality, and the Observer*, Keshav will step into global intellectual battlefields.

He will no longer walk alone. He will walk into the storm—where physics, philosophy, politics, and psychology all demand answers.

Not as a guru. Not as a rebel. But as a man with only one purpose:

To awaken the intelligence that sleeps beneath every belief.

The monk has spoken. Now, the world will respond.

Acknowledgements

This book was never planned. It was born—quietly, steadily—from a series of questions that refused to stay silent.

To everyone who has ever paused mid-sentence, mid-life, or mid-success to ask, *"Is this all there is?"*—this book was written in your name.

I offer my deepest gratitude to the **invisible forces of alignment**—those moments of stillness, contradiction, and collapse that shaped the path far more than applause or certainty ever could.

To the **scientific community**—for giving me a foundation strong enough to walk away from, and for teaching me how to ask the "how" so that I could eventually seek the "who."

To the **sages, seekers, and skeptics**—from ancient texts to modern thinkers—who whispered (or shouted) across time, reminding us that no system, no theory, no doctrine is above inquiry.

To those close to me—family, friends, and companions in both silence and chaos—who witnessed the shift and chose to support rather than understand. Your patience is part of this journey.

To my **students, critics, and fellow observers**—thank you for your questions. Especially the uncomfortable ones. You've kept me honest.

To the **characters who shaped this narrative**, especially Keshav—the monk who was once a man. Though fictional in form, you are very real in essence. You live inside every one of us who dares to break free.

To the readers of this first volume—thank you for being willing to start at the beginning, knowing full well there is no final destination—only unfolding.

And lastly, to **Intelligence itself**—that ever-present, ripple-born awareness that guides, questions, and awakens—thank you for letting me be a temporary vessel for your voice.

This is just the first step. The tour has begun. And so has the truth.

— *With alignment and intent,*

Sandeep Chavan

About the Author

Sandeep Chavan is an independent thinker, educator, and researcher who walks between worlds—of science and spirit, logic and intuition, silence and rebellion. With a background in engineering and physics, he spent years decoding complex systems, only to realize that the most urgent truths lie beyond equations.

He is the founder of **Universal Energy Dynamics (UED)**—a paradigm-shifting framework that challenges conventional science by replacing force and motion with ripple-based alignment and field intelligence.

A teacher by passion and a seeker by nature, Sandeep's work blends rigorous inquiry with fearless questioning. He writes not to convince, but to crack open perspectives. His books cross genres—from physics to philosophy, from war studies to personal transformation—with one common thread: **the search for deeper alignment.**

Sandeep lives simply, teaches profoundly, and believes that truth doesn't need protection—only perception.

With *The Monk on World Tour* series, he invites readers into a journey not of belief, but of awakening—where intelligence replaces ideology, and freedom begins with asking the right questions.

About the Series

A **Monk on World Tour** is not just a book series—it is a living dialogue between science, spirituality, and society.

Told through the eyes of Keshav—a brilliant scientist who walks away from fame, certainty, and comfort to pursue a deeper truth—this series invites readers into a global expedition of awakening. But Keshav is no ordinary monk. He wears no robes. He offers no mantras. His mission is not to preach, but to provoke thought, dissolve illusions, and awaken **intelligence in motion**.

What Makes This Series Unique?

- **Narrated in the Monk's Voice** – Every chapter unfolds through his eyes, thoughts, and struggles.
- **No heavy theories** – Complex ideas are made accessible through conversations, debates, and raw human experience.
- **Real-World Characters** – Scientists, spiritual leaders, business moguls, atheists, students, fools—and everyone in between.
- **Unfiltered Debates** – No sugarcoating, no

avoidance of controversy. Questions are asked fearlessly and answered with depth.

- **Safe but Bold** – Religious, scientific, and philosophical boundaries are challenged—without disrespect, but with daring honesty.
- **Humor & Humanity** – Even the deepest truths are delivered with humility, wit, and emotional realism.

The Journey Across 9 Volumes:

1. **The Making of the Monk:** Keshav's transformation—from an acclaimed scientist to a monk who no longer follows systems, but **awakens them**.
2. **Time, Reality, and the Observer:** Global debates on what time really is—and how reality depends not on clocks, but on perception.
3. **The Free Will Paradox:** A sharp exploration into destiny, choice, karma, and whether we are truly free—or just flowing in an unseen current.
4. **The Science of Ignorance:** A philosophical face-off where Keshav argues that ignorance is not a flaw, but the **engine of evolution**.
5. **The True Nature of God:** Beyond religion, Keshav introduces God as the **intent of intelligence in motion**, challenging both faith and atheism.
6. **Faith: Pure, Blind, and Polluted:** An examination of how faith was distorted—and how it can be restored, not as belief, but as alignment.
7. **God, Consciousness, and Universal Intelligence:**

The deepest of all debates—what really animates us? Is consciousness a chemical accident or cosmic intelligence?

8. **The Future of Humanity:** Keshav meets tech leaders and thinkers at the edge of civilization. Can intelligence survive artificial imitation?
9. **Closing the Loop: The Ultimate Global Seminar:** The Monk returns to the world stage, connecting every theme, question, and realization into one final message.

Why This Series Matters

In an age of noise, belief wars, and intellectual ego, Sandeep Chavan's *A Monk on World Tour* does something radical:

It makes you **feel** intelligence. It makes you **question** everything you thought you knew. And above all, it offers you not a philosophy, but a **mirror**.

This is the series for thinkers, seekers, rebels, scientists, skeptics, and anyone who refuses to settle for shallow truths.

Welcome to the journey. The Monk has arrived. And the tour has just begun.

Don't miss out!

Visit the website below and you can sign up to receive emails whenever Sandeep Chavan publishes a new book. There's no charge and no obligation.

https://books2read.com/r/B-A-EPGPC-OVITG

Connecting independent readers to independent writers.

Did you love *The Making of the Monk*? Then you should read *Chavan's Declaration*[1] by Sandeep Chavan!

[2]

Chavan's Declaration*What if everything you've ever believed—about science, spirituality, and even your own self—is just a ripple on a deeper, structured field?*

Chavan's Declaration is not merely a book. It is a philosophical earthquake. A bold unification of ancient wisdom and radical science, it introduces a revolutionary framework: **Universal Energy Dynamics (UED)**—a ripple-based understanding of energy, consciousness, and reality itself.

---

1. https://books2read.com/u/47pp2L

2. https://books2read.com/u/47pp2L

This is not about rejecting science or worshipping mysticism. It's about decoding what both have been trying to express all along—using different waveforms. Beneath matter and mind lies **Structured Vacuum Energy (SVE)**—a still field where all motion, form, and perception are ripple resolutions.

From gravity to God, from death to thought, from time to truth—this book dismantles outdated assumptions and replaces them with ripple logic:

The Chavan Ripple Field Equation (CRFE)—a proposed upgrade to Einstein's Field Equations**Mind, soul, and consciousness**—redefined as ripple behaviors, not metaphysical mysteries**Time**—not a flowing dimension, but a ripple consequence of causal tensions**Black holes**—not pulling matter, but resolving field imbalance**Religion, science, and scriptures**—recast as attempts to describe the same underlying alignment**AI vs. Human Awareness**—a ripple-level difference, not a computational one**And the final declaration:** Truth is not fixed—it's functional.

Whether you're a seeker, a skeptic, a teacher, or a physicist—this book doesn't ask you to agree. It invites you to **align**. To see not with belief, but with coherence.

Chavan's Declaration is the culminating volume of *The War of Thoughts* series—a philosophical rebellion against dogma, and an intellectual invitation to surf the ripple of truth.

The universe isn't broken. Only our explanations are.Dare to update them.

Read more at https://www.sandeepjchavan.com/.

Also by Sandeep Chavan

A Monk on World Tour

The Making of the Monk

Chavan's UED research

CMGED Research Report

Chavan's Cosmic Continuum (C^3)

Frequency

Ripples of Reality

Reality Realigned

The Great Realignment

I – The Cosmic Leverage

Chavan's Cosmic Theory of Relativity (CTR)

Chavan's Theory of Light

Feeling Like a Physicist

TIME – The New Definition

$E \neq mc^2$: The Collapse of Classical Assumptions

The Chavan Codex

Chavan's Atomic Model

Chavan's Theory of Everything

Echoes of War: The Russia-Ukraine Conflict and Its Global Ripples

Shattered Lives

The War Economy

Borders Redrawn

Arms and Allies

Collateral Damage

Justice in the Ashes

Somewhere in the Darkness of the Cosmos

The Blueprint of the Universe

The Cosmic Breath

From the Edge of the Universe

The Eternal Sage Chronicles

The Sage and the Sovereign

The Family Pillar

The People's Flame

The Warrior's Soul

The Bridging Mentor

The Guiding Light

The Exodus

The Last Cycle: 2075

The Fundamental Theory of Life & Existence

Energy in Action: The Soul Demystified

The Active Mind

Energy in Connection

Death: The Disconnection

Subconsciousness: The Chosen Disconnection

Desires: The Engine of Life

The Restoration of Human Knowledge

Restoring Bhagavad Gita

The War of Thoughts

Cracks in the Known

The Human Game

The Divine Dilemma

The Original Intent

The Answers

Ultra Modern Life Philosophy

The Duality of Laughter and Tears

Take It Easy

Compare to Accept

Standalone

It's Not AI, It's AHI - Amplified Human Intelligence
The Decision Paradox
The Four Sapiens
Malicious Script of Indian Polity
The IIT Legacy & Global Impact
The IIT Legacy & Global Impact
You are just a Version of Your Original
Win the Game You Didn't Choose
Always Keep Your Bags Packed
Reveal to Shield - Learn the Game of Facades
Engage Beyond Elections
Culture, Identity & Change: The Evolution of Indian Society
The Triangular Dynamics
Never Trust Without Doubt
If You Can't Change the Country, Leave the Country
Live Fearless with Fear
India: The Investment Magnet
Realignment
Powering a Billion Dreams
Universal Skill
Extension: From Brush to Algorithm
Human Side of Einstein
The Future Knows No Lines
Let's Bring the Generations to the Table
Unified Energy Dynamics
A Beat That Never Dies
Strategic Horizons

Right From Scratch
The Forgotten Innovators
Physics Fixing
The Savior Illusion
The Selfish Side
The Age of the Weakest Sapiens
Neural Wars
The Absorption: There Is No Death
Strange Therapist
Bubbles: Become a Drop
Beyond the Blue
Half-Scholar
The Law of Truth
Failure
Philosophy 2.0: A Structural Manifesto
Chaos: A Gentle Guide to the Unresolvable
Self-Help: New Syllabus
Rippling
Game: How to Play the Share Market
Out of Syllabus
Saffronized: The Code Behind the Crown
The Anger
Moksha: The Liberation
Chavan vs Einstein: The Universe Was Never Curved
Rocket Science: The Ripple Gambling
The Cosmos: The Still Universe
The Taste of Silence
Chavan's Post-Materialist Human Model
10-Second Rule
Chavan vs Hawking
It was Never a Wave Signal

Quantum Courtroom: The Trial of a Century
Chavan–Heisenberg Ripple Uncertainty Principle
The Great Disruption
Dark ≠ Hidden
Breath
Calculus of Life
The Earth Doesn't Move
Black Hole Don't Exist
Chavan vs Kant
Naked Truth: Everything and Else
The Moon: Co-Author of Life's Script
World War-3
Aham Brahmasmi
Chavan's Declaration

Watch for more at https://www.sandeepjchavan.com/.